Minority Shareholder Monitoring and German Corporate Governance

Corporate Finance and Governance
Herausgegeben von Dirk Schiereck

Band 13

Christian Thamm

Minority Shareholder Monitoring and German Corporate Governance

Empirical Evidence and Value Effects

PL ACADEMIC RESEARCH

Bibliographic Information published by the Deutsche Nationalbibliothek
The Deutsche Nationalbibliothek lists this publication in the Deutsche Nationalbibliografie; detailed bibliographic data is available in the internet at http://dnb.d-nb.de.

Zugl.: Darmstadt, Techn. Univ., Diss., 2013

Cover Design:
© Olaf Glöckler, Atelier Platen, Friedberg

Library of Congress Cataloging-in-Publication Data

Thamm, Christian, 1980-
 Minority shareholder monitoring and German corporate governance : empirical evidence and value effects / Christian Thamm.
  pages cm. — (Corporate finance and governance ; band 13)
 Includes bibliographical references.
 ISBN 978-3-631-64086-9
 1. Minority stockholders—Germany. 2. Corporate governance—Germany. 3. Corporations—Valuation—Germany. I. Title.
 HG4156.T49 2013
 338.6—dc23
                                        2013008200

D 17
ISSN 1869-537X
ISBN 978-3-631-64086-9
© Peter Lang GmbH
Internationaler Verlag der Wissenschaften
Frankfurt am Main 2013
All rights reserved.
PL Academic Research is an Imprint of Peter Lang GmbH.

Peter Lang – Frankfurt am Main · Berlin · Bruxelles · New York · Oxford · Wien · Warszawa

All parts of this publication are protected by copyright. Any utilisation outside the strict limits of the copyright law, without the permission of the publisher, is forbidden and liable to prosecution. This applies in particular to reproductions, translations, microfilming, and storage and processing in electronic retrieval systems.

www.peterlang.de

Preface

How changes in a firm's ownership structure influence the corporate governance and, consequently, the corporate valuation has been widely discussed in the finance literature for decades. However, many important questions raised in this discussion remain unanswered today – in particular with regard to active shareholders like single active private investors and hedge funds. The core market for these financial investors is the U.S. market, which is the most mature and well-established market with a long-run history and a large number of very experienced market participants. In contrast, the German market is younger and smaller but has been characterized by a strengthening of shareholder rights and a sharp increase in transaction volumes during recent years. Whether empirical evidence for the U.S. market can be transferred to the German institutional setting is highly questionable because the political, legal and financial environment in Continental Europe deviates remarkably from the Anglo-Saxon countries.

In his thesis, Christian Thamm sets out to analyze whether active investors generate shareholder value when they acquire minority stakes in German corporations and influence the corporate governance. This is not only a remarkable endeavour, as Mr. Thamm uses hand-collected, unique German data to present state-of-the-art analyses, which are competitive and meet highest international standards. The thesis on hand carefully identifies and addresses open research questions related to the shareholder activism in German corporations. Its primary objective was to identify value consequences of shareholder activism and corporate governance changes in Germany analyzing stock price data.

Mr. Thamm fully achieves the objectives of his dissertation. The analyses contain many intriguing and surprising results, which make this thesis a more than interesting read, that I highly recommend to corporate finance researchers and investor relations practitioners. I wish for the dissertation its due wide diffusion in corporate finance and investor relations research.

Prof. Dr. Dirk Schiereck

Acknowledgements

First and foremost, I want to thank Prof. Dr. Dirk Schiereck for the cordial welcome in Darmstadt and the good organization of the doctoral programme. I appreciate the constant feedback on my research that I have received over the past two and a half years. This was very, very helpful.

The support from my colleagues at the Chair of Corporate Finance at Tech University Darmstadt was great. Christoph Böhm, Anit Deb, Daniel Maul, Rahmit Mehta, Malte Raudszus and Julian Trillig provided useful advice. Thanks goes to Matlab programming experts Christian Babl and Steffen Meinshausen. I enjoyed the discussions on shareholder activism with Christian Hertrich, Mark Mietzner, Christian Rauch and Maximilan Stadler.

Goethe University Frankfurt provided free access to their databases and journals library and I am thankful for that. I also appreciate the support from WAI Wirtschaftsanalysen- und Informations GmbH with respect to providing free access to their database. This was an invaluable source of historic corporate governance information on German corporations.

I ran regressions and some of the statistical tests using the Gnu Regression, Econometrics and Time-series Library. I would like to say thank you to the programming team.

The research conferences that I attended added great value to my work, especially the 2011 Doktorandenseminar at the Deutsche Bundesbank facility in Eltville, Germany.

Needless to say, all this would not have been possible without the support from my family and my friends.

I appreciate it.

Christian Thamm

Table of Contents

Preface V

Acknowledgements VI

Table of Contents VII

List of Figures X

List of Tables XI

List of Abbreviations XIII

List of Symbols XV

I.   Introduction 1

I.1.   Preface 1

I.2.   Definition of Shareholder Activism and Corporate Governance 2

I.3.   Theoretical Framework 6

I.4.   Research on Shareholder Activism in Germany 7

I.5.   Research on Shareholder Activism in the U.S. 13

I.6.   Development of Hypotheses 14

I.7.   Remarks on the Collection of Data 15

I.8.   Presentation of Studies and Contribution to Literature 19

    I.8.1.   First study: Shareholder Activism in Germany – An Empirical Review 19

    I.8.2.   Second Study: Activist Shareholders, Abnormal Returns, and the German Aufsichtsrat 19

    I.8.3.   Third Study: Weak Shareholder Rights? – A Case Study of Cevian Capital's Investment in Demag Cranes AG 23

II. Shareholder Activism in Germany – An Empirical Review 25

II.1. Introduction 26

II.2. Literature Overview 27

II.3. Data Collection 31

II.4. Descriptive Statistics: Target Firm, Stake Size and Activist Shareholders 35

II.5. Shareholder Activism and its Effects 40

II.6. Summary of Results 48

III. Activist Shareholders, Abnormal Returns, and the German Aufsichtsrat 51

III.1. Introduction 52

III.2. Three Cases of Shareholder Activism 53

III.3. Literature Background, Corporate Governance Framework and Research Hypotheses 54

III.4. Data and Methodology 57

III.5. Results 60

    III.5.1. Announcement Effects 60

    III.5.2. NewBET Analysis 63

    III.5.3. Determinants of Abnormal Returns 68

III.6. Conclusion 71

IV. Weak Shareholder Rights? – A Case Study of Cevian Capital's Investment in Demag Cranes AG 73

IV.1. Introduction 74

IV.2.	German Supervisory Board and U.S. Board of Directors 77
IV.3.	Transaction Partners.. 81
IV.4.	Investment by Cevian Capital and Takeover by Terex Corporation .. 85
IV.4.1.	Investment by Cevian Capital 85
IV.4.2.	Supervisory Board Representation 86
IV.4.3.	Takeover by Terex Corporation.................................... 87
IV.4.4.	Lessons Learned From the Demag Case....................... 88
IV.5.	Cevian Capital's Follow-up Investment in Bilfinger Berger SE........ 89
IV.6.	Shareholder Activists on German Supervisory Boards 90
IV.7.	Conclusion.. 94

V. Summary of Key Research Findings 95

References 97

List of Figures

Figure I.1. Stock Market Investment Philosophies 4

Figure I.2. Shareholder Structure of German Publicly-Traded Corporations 1963-83 8

Figure I.3. Quarterly Distribution of Events 18

Figure II.4. Daily Trading Volume of Target Firms 41

Figure III.5. Relationship Between Timing and CAR 66

Figure VI.6. Annotated Share Price Graph of Demag Cranes AG (ISIN DE000DCAG010) Between 01 January 2010 and 31 July 2011 84

List of Tables

Table I.1. Definitions of Shareholder Activism 2

Table I.2. Research Methodologies and Results of Studies on Shareholder Activism 10

Table I.3. Possible Definitions of Activist Event and Evaluation for Event Study Purpose 16

Table I.4. Matrix of Pearson Sample Correlation Coefficients 22

Table II.5. Overview of Empirical Studies on Shareholder Activism in Germany 30

Table II.6. Comparison of Disclosure Requirements in the U.S. and in German 32

Table II.7. Introduction of Activity Levels 35

Table II.8. Percentage Size of Activist Stakes 36

Table II.9. Value of the Activist Stakes in EUR Million 37

Table II.10. General Classification of Potentially Activist Shareholders 38

Table II.11. Timing of Activist Investments 39

Table II.12. Change in Annual Meeting Attendance Rates 43

Table II.13. Simultaneous Investments by Multiple Activists 44

Table II.14. Possible Effects of low Attendance Rates in the Sample 46

Table II.15. Major Corporate Events at Sample Target Firms. 47

Table III.16. Sample Description 59

Table III.17. Announcement Effects on Different Activity Levels 61

Table III.18. Announcement Effects of Non-Activist Investments 62

Table III.19.	Announcement Effects and NewBET	64
Table III.20.	Difference-in-Means-Test	67
Table III.21.	Results of Ordinary Least Squares Regression	69
Table IV.22.	Comparison of U.S. Board of Directors and German Supervisory Board	78
Table IV.23.	Company Description and Market Information	82
Table IV.24.	Shareholder Activists on German Supervisory Boards	92

List of Abbreviations

AktG	Aktiengesetz
AG	Aktiengesellschaft
AGM	annual general meeting
approx.	Approximately
BaFin	Bundesanstalt fuer Finanzdienstleistungsaufsicht
BAWe	Bundesaufsichtsamt fuer den Wertpapierhandel
BZ	Boersen-Zeitung
CAPM	Capital Asset Pricing Model
CAR	Cumulative abnormal return
CEO	Chief executive officer
Co.	Corporation
CSR	Corporate social responsibility
DAX	Deutscher Aktienindex
DCGK	Deutscher Corporate Governance Kodex
DGCL	Delaware General Corporation Law
EBIT	Earnings before interest and tax
et al.	Et alii
EUR	Euro
EURm	Euro millions
e.V.	Eingetragener Verein
GmbH	Gesellschaft mit beschraenkter Haftung
IPO	Initial public offering
ISIN	International Securities Identification Number
KKR	Kohlberg, Kravis and Roberts
LLP	Limited liability partnership
Log	Logarithm
LSE	London Stock Exchange
Ltd.	Company with limited liability
Market cap	Stock market capitalisation of a listed company
Max.	Maximum
Mgmt.	Management
NewBET	New Supervisory Board Election Timing
N.Y.	New York
obs.	Observations
p-value	Probability of obtaining a certain test statistic
ROE	Return on equity
SA	Sociedad Anónima

SE	Societas Europaea
SEC	U.S. Securities and Exchange Commission
S&P	Standard & Poors
TCI	The Children's Investment Fund
TIAA	Teachers' Insurance and Annuity Association
U.S.	United States of America
USD	United States Dollar
WpHG	Wertpapierhandelsgesetz
WpÜG	Wertpapiererwerbs- und Übernahmegesetz
II ZR	Committee at the German national supreme court

List of Symbols

$\hat{\alpha}_i$	CAPM: estimated alpha
$\hat{\beta}_i$	CAPM: estimated beta
b_0	Regression: constant
b_i	Regression: coefficient
CAR	Event study: Cumulative abnormal return
ε_{it}	Event study: abnormal return of firm i on day t
i	Event
i	Firm
i	Observation
N	Number of observations
r	Pearson's sample correlation coefficient
R_i	CAPM: stock return
R_m	CAPM: market return
$\sigma_{\bar{x}_i - \bar{x}_j}$	Variability of the differences between two sample means
t	Event study: day
t	Testing: test statistic of standard T-test
x_i	Regression: control variable
$\bar{X}_i$	Sample mean
y	Regression: dependent variable
z	Testing: Boehmer-Poulsen-Musumeci test statistic
z	Testing: Standardised Wilcoxon test statistic

I. Introduction
I.1. Preface

Monitoring by minority shareholders is a topic that is receiving more attention in Germany – even though shareholder activism is not a completely new phenomenon. A description of an early case of shareholder activism in Germany can be found in the journals Zeitschrift für Wirtschafts- und Bankrecht (1961) and Betriebs-Berater (1962). Even though the documentation neither states the name of the activist nor the corporation it does reveal some interesting facts on this early incident: in 1957, a minority shareholder of a German corporation at the annual meeting of that corporation asked for additional information regarding executive compensation and he also called for a higher dividend, deeming the management's dividend proposal too low. The management denied both the disclosure of the requested information and paying a higher dividend. The activist also urged the corporation to buy back his shares at a premium. The German national supreme court Bundesgerichtshof later ruled that the shareholder's claims both in terms of information rights and the higher dividend were legitimate (case II ZR 4/60 ruled on 23 November 1961). The corporation who acted as defendant in the case criticized the shareholder as overly aggressive and being driven by speculation purposes.

An early and well-known proponent of minority shareholders' rights was Erich Nold, originally a coal trader from Darmstadt, Germany. He became received attention as a BMW shareholder fending off a takeover bid by rival car maker Daimler-Benz in 1959 which he and other minority shareholders deemed too low. This is an example of how small shareholders have shaped the face of corporate Germany.

Iber (1985) in his study finds that between 1963 and 1983 the circumstances for minority shareholder activism in Germany were less attractive and less favourable. These circumstances prevailed for a couple of years.

Over the past two decades the environment for activist shareholders in Germany became more amenable ((Schaefer, 2007), (Goergen, Manjon and Renneboog, 2008)). Therefore the purpose of this study is to provide an in-depth analysis of the new minority shareholder activism in Germany by presenting three studies. The first study gathers empirical evidence on shareholder activism in Germany. The second study measures and analyzes abnormal stock returns surrounding the disclosure of an activist stake. The third study illustrates a recent case of minority shareholder activism and sheds more light on the legal environment for activist shareholders in Germany.

The introductory part of this dissertation proceeds as follows. In the next section the concept of shareholder activism and the meaning of corporate governance in the context of this study will be discussed. This is followed by an introduction to the theoretical framework of the analysis. The two subsequent sections are literature reviews of shareholder activism and related topics in Germany, and in the U.S. The chapter then continues with the development of hypotheses and some general, but important, considerations on the process of data collection. The last part of the introduction is a preview of the three research papers presented. It includes some additional information on the research methodologies applied and highlights the contributions to literature of each study.

On the basis of the several hundreds of events reviewed and the many analyses conducted, the German stock market appears to be an efficient market in the sense of Fama (1970). There was no evidence found that capital market participants are not well-informed or that their behaviour is similar to that of irrational stock market traders. This is an important assumption in conducting this study, especially the various analyses of the value effects.

I.2. Definition of Shareholder Activism and Corporate Governance

There is no single definition of shareholder activism and there is no database that provides a comprehensive list of activist shareholders or activist events. Therefore, Table I.1. provides an overview of how recent studies whose research is related to corporate governance topics define shareholder activism.

Table I.1. Definitions of Shareholder Activism; The definitions by Prigge and Steenbock and Stadler have been translated from German

Author	Definition
Bethel (1998)	The announced intention of influencing firm policies.
Prigge and Steenbock (2002)	Every corporate governance measure initiated by a shareholder.
Gillan and Starks (2007)	Investors who are dissatisfied with some aspect of a company's management or operations and try to bring about change within the company without a change in control.
Stadler (2010)	Shareholders who are unsatisfied with the progress of the firm try to influence and improve the strategy through communication with, and pressure on, the management but without the aim of acquiring the whole firm.

In the spirit of Bethel (1998), a *potential* activist shareholder is a known investor who *might* announce an intention of influencing firm policies in the future. The concept of *potential shareholder activist* will be used throughout the analysis. In addition, events will be categorized by the level of activism.

Tirole (2006) presents a more formal definition of "active monitoring". Active monitoring is associated with either formal or real control. A majority of votes on the board or a majority of votes in the annual meeting are prerequisites for formal control. According to Tirole, real control exists when minority shareholders succeed in persuading a majority of the board or a majority of fellow shareholders to go along with a given policy.

Applying the concept of Tirole, this study would were to be classified as an analysis of cases of active monitoring through real control. As will be seen later, activist shareholders do sometimes obtain board seats, but minority stakes usually do not result in formal control.

Figure I.1. describes different investment philosophies. Understanding these investment philosophies and their implications for the markets is important. Some of these investment styles contribute to the existence of agency problems and, most importantly, not all investors apply strategies of close monitoring. Investors, that apply the investment style of indexing, usually do not analyze stocks and they do not engage in monitoring activities. These investors simply follow stock index compositions such as the Dow Jones or the S&P 500. Stock-picking investors conduct financial analysis prior to investing. They choose to not actively engage in monitoring and remain passive investors. Structural corporate governance changes may be promoted by investors managing large diversified stock portfolios. Some U.S.-pension funds actively engage in monitoring through structural corporate governance changes. Structural corporate governance changes can be implemented, for example, with respect to board composition, policies for capital increases or major strategic decisions, such as large acquisitions. Bassen (2002) provides more detail and categorizes these policies. Implementing such rules does not necessarily require an in-depth understanding of a corporation's business model but it rather aims at creating certain standards and procedures of good corporate governance. *Shareholder activism* aims at single firms. It requires both, in-depth analysis and understanding of a firm's business beyond mere financial analysis, and active monitoring, for example through supervisory board representation.

Figure I.1. Stock Market Investment Philosophies; Note: in the literature Indexing is referred to as "passive" investing while Stock picking is sometimes referred to as "active" investing. This is not to be confused with activist investing

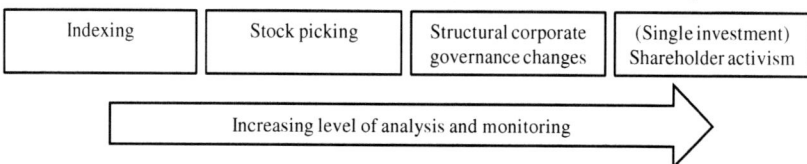

| Indexing | Stock picking | Structural corporate governance changes | (Single investment) Shareholder activism |

Increasing level of analysis and monitoring ⟶

At the centre of attention of the analysis is the observed behaviour of the activist shareholders and not their identity, especially not the self-definition of the investor as an activist or non-activist or any other certain self-definition of the investor. Some investors label themselves as non-activists even though they do apply activist strategies and vice versa. An example, why the observed behaviour and not any kind of self-definition should be at the centre of attention, is the case of Blackstone Group's investment in Deutsche Telekom AG in 2005. Blackstone acquired 4.5 percent of the shares without any intention of acquiring the whole firm then or in the future. Blackstone achieved supervisory board representation through taking a seat on the supervisory board of Deutsche Telekom AG the same year. This means Blackstone acquired a minority stake, did not have any intentions to acquire the whole firm and obtained supervisory board representation. Supervisory board representation offers the possibility to substantially influence firm policies and can, therefore, be viewed as the highest form of shareholder activism (Schaefer, 2007). Blackstone Group, in its own view, has always followed a private equity strategy. Its strategy as a private equity firm is to acquire whole companies and not minority stakes with the aim of engaging in minority shareholder activism. This illustrates how following the self-definition of an investor might result in drawing a wrong conclusion about the investor's strategy since investors tend to follow opportunistic investment strategies.

What is the objective of shareholder activists? The modern corporation is characterized by the separation of ownership and control. Large corporations are not run by their owners, the shareholders, but by hired managers. A hired manager may run the company in the best interest of the shareholders (stewardship theory) but he might as well take decisions that are in his own interest or simply not in the best interest of the corporation (agency theory). Strong empirical evidence for agency conflicts exists as will be shown later during the analysis.

How do activist shareholders create value? Brav, Jiang, Partnoy, and Thomas (2008) in their widely cited paper on hedge fund activism in the U.S. provide an example on how lowering disproportionate executive compensation increases firm value:

"If all of the top executives combined are paid USD 5 million less for 5 years post intervention due to activism, and this value goes to shareholders (assuming away tax issues, etc.), then the present value of such an income stream is on the order of magnitude of USD 20 million (using a 10% discount rate), which is a significant portion of the market capitalization of a typical targeted company (...)".

Assuming a target firm had a market capitalisation of USD 500 million, this decrease in executive compensation would immediately increase firm value by 4 percent and accordingly, the value of the activist's stake.

There is generally speaking a lower level of executive compensation in Germany. Shareholder activists in Germany may focus on different aspects of agency conflicts beyond executive compensation. The intention of their actions may include: gaining a supervisory board seat for general monitoring purposes, asking a member of the management board or the supervisory board to stand down, publicly criticizing the target firm's corporate governance, influencing or forcing certain strategic decisions, calling for a higher dividend or forcing the target firm into a takeover. There are several possibilities for communicating demands besides direct communication. This includes communications via newspapers, investor letters issued by the activist fund or, as well, the annual general meeting of the target company. This all assumes that previous quiet communications between activist and target firm have not led to a solution of the issue or were, for some reason, not intended by the activist.

In general linguistic usage, but also in academic linguistic usage, shareholder activism is sometimes used in the sense of corporate social responsibility activism. The topic of corporate social responsibility activism or CSR activism (see for example Doh and Guay (2006)) is not aimed at overcoming agency problems between owners and managers of the firm in the sense of Jensen and Meckling (1976) and therefore not covered in this study.

Shleifer and Vishny (1997) in their survey of corporate governance put forward:

"Corporate Governance deals with the ways in which suppliers of finance to corporations assure themselves of getting a return on their investment. (...) How do suppliers of finance control managers? (...) After all, they [the suppliers] part with their money, and have little to contribute to the enterprise afterward."

Corporate governance in the context of this study is defined in the spirit of Shleifer and Vishny. The focus is on the control of the management board and supervisory board through blockholders – in the case of this analysis, minority blockholders. Schiereck and Steiger (2001) point out that especially blockholders of medium-sized firms have a strong interest in improving a firm's governance and thereby increasing the value of their holding. Selling the block on the open market, in case of dissatisfaction with the firm's corporate governance, would most likely result in a disproportionate price decline.

The day-to-day management of a German corporation is assigned to the management board (Vorstand) while the supervisory board (Aufsichtsrat) shall monitor the management board as stipulated by German corporation law. Nevertheless, agency conflicts can still arise and they do arise. This will be discussed in the next section.

I.3. Theoretical Framework

The Principal-Agent-Theory is the framework most frequently applied to academic research on corporate governance topics. This theory focuses on the issue of the separation of ownership and control.

Probably the first scholar to dedicate his research to this topic was Adam Smith (1776). In his seminal work "An Inquiry into the Nature and Causes of the Wealth of Nations" Smith studied the division of labour. He remarked that

"The directors of such [joint-stock] companies, however, being the managers rather of other people's money than of their own, it cannot be well expected, that they should watch over it with the same anxious vigilance with which the partners in a private copartnery frequently watch over their own (...)."

Other researchers (Berle and Means, 1932) followed with similar explanations but putting the whole argument into a more modern context:

"The property owner who invests in a modern corporation so far surrenders his wealth to those in control of the corporation that he has exchanged the position of independent owner for one in which he may become merely recipient of the wages of capital (...). [Such owners] have surrendered the right that the corporation should be operated in their sole interest (...)."

Fama and Jensen (1983) further developed the discussion pointing out that

"Our concern (...) is with the organizational forms characterized by separation of decision management from residual risk bearing – what the literature on open corporations calls, somewhat imprecisely, separation of ownership and control."

Jensen and Meckling (1976) seven years earlier in their seminal research paper *"Theory of the Firm: Managerial Behaviour, Agency Costs and Ownership Structure"* integrated elements from various theories such as the theory of agency, the theory of property rights and the theory of finance to develop a new theory of the ownership structure of the firm. Their study also defined the concept of agency costs.

Agency theory predicts that agency costs do exist and that they reduce the value of the firm. The theory provides tools to reduce agency costs, for example by monitoring of the management. Throughout the literature many different examples for agency costs are discussed. These costs can arise from such things as perquisites for management (Yermack, 2006), entrenched boards or arrangements that protect management from removal (Bebchuk and Cohen, 2005) and entrenching investments (Shleifer and Vishny, 1989). Yermack in his study shows how U.S. firms, whose CEOs personally use corporate aircraft, underperform market benchmarks. In this context the personal use of corporate aircraft can be interpreted as ineffective cost control.

Agency costs have also been discussed in relation with mergers and acquisitions. Homberg and Osterloh (2010) provide a comprehensive overview of empirical findings. According to Homberg and Osterloh, a large number of takeovers economically fail given the overconfidence of the decision maker. The decision maker can easily overestimates his abilities in raising potential synergies. Therefore the decision-maker is likely to make investment decisions that destroy shareholder wealth. Prominent examples for acquisitions that destroyed shareholder wealth on a large scale are Deutsche Telekom AG's investment in T-Mobile USA, Daimler AG's acquisition of Chrysler Corporation and the attempt of several acquirers, including Fortis Bank, to take over Dutch bank ABN Amro. The latter ultimately contributed to the near-bankruptcy of Fortis Bank and the firm was ultimately acquired by BNP Paribas. Wenger (2008) documents how the share price of Daimler AG has underperformed benchmark stock BMW AG given a series of unfavourable management decisions. Shleifer, Morck and Vishny (1990), Moeller, Schlingemann and Stulz (2004) and Moeller, Schlingemann and Stulz (2005) all find negative abnormal announcement stock returns to acquiring firms under certain circumstances, for example, for frequent acquirers or for large acquirers.

I.4. Research on Shareholder Activism in Germany

In 1985, Bernhard Iber published a research study on the development of the shareholder structure in Germany between 1963 and 1983. During this time period there was a substantial increase in equity ownership by banks, insurance companies and corporations as shown in Figure I.2. Cross-shareholdings and cross-directorships among German corporations were all but infrequent. This phenomenon was by many labelled as the "Deutschland AG" ("Germany Incorporated").

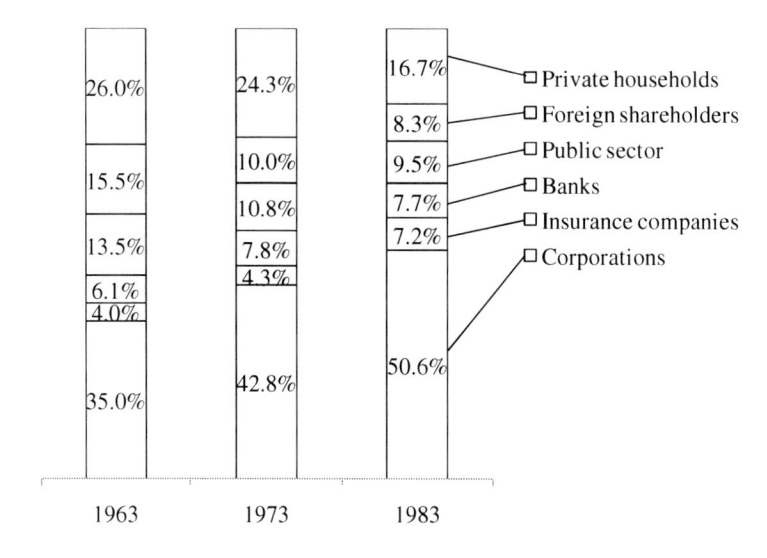

[Chart showing shareholder structure across 1963, 1973, 1983]

1963: 26.0%, 15.5%, 13.5%, 6.1%, 4.0%, 35.0%
1973: 24.3%, 10.0%, 10.8%, 7.8%, 4.3%, 42.8%
1983: 16.7%, 8.3%, 9.5%, 7.7%, 7.2%, 50.6%

Legend: Private households, Foreign shareholders, Public sector, Banks, Insurance companies, Corporations

Figure I.2. Shareholder Structure of German Publicly-Traded Corporations 1963-83; Source: Iber (1985)

Iber concluded that the German stock market especially for small, private shareholders and institutional investors – both can be seen as the predecessors of today's minority activist shareholders – had lost some of its attractiveness in terms of the number of stock-market listed companies and given the constant decrease in free float. The long-term nature of the investments made by banks, insurance companies and other corporations contributed to this development.

The environment for shareholder activism by minority shareholders has changed fundamentally since then. This was due to several factors, such as the unwinding of corporate ownership, that was described by Iber (1985) and the declining influence of banks and bank directorships on supervisory boards (Dittmann, Maug and Schneider, 2010). It can also be attributed to changes in the German corporate governance system and major regulatory initiatives for transparency and accountability (Goergen et al., 2008).

A few studies have investigated minority shareholder activity in Germany in the past. These studies will be introduced now.

Jenkinson and Ljungqvist's (2001) study focused on hostile stakes in conjunction with takeover activity in Germany between 1987 and 1994. The study covered an observation period that can be described as an environment of departures from the one-share-one-vote principle, potentially limited share transferability, possible caps on voting rights and listings of only non-voting shares. Still Jenkinson and Ljungqvist's article shows how *"the German system of corporate control and governance is both more active and more hostile than has previously been suggested"*.

The first study to investigate minority shareholder activism in Germany using the definition of a *"corporate raider"* as *"a minority shareholder who is expected to force changes in the target firm's corporate policies, based on his reputation for annoying incumbent management"* was Croci (2007). Croci's view on corporate raiders in Europe stated that these investors have superior stock-picking ability but do not improve firms' corporate governance through their monitoring. His study did not focus on Germany exclusively but included several European countries with 17 observations from Germany. The observation period ranged from 1990 to 2001. It was therefore not identical to that of Jenkinson and Ljungqvist, but similar.

The first study of voting control in German corporations to include disclosures below the 25 percent threshold based on WpHG-filings was Becht and Boehmer (2003). The WpHG is the German securities trading act (Wertpapierhandelsgesetz). The authors investigated the level of ownership concentration and concluded a high level of ownership concentration in Germany as of 1996. What Becht and Boehmer called a "substantial difference" in method and data used, was actually a turning point for studies of ownership and voting power of German corporations.

Mietzner and Schweizer (2007) were then the first authors to conduct an event study of shareholder activism by new blockholders on the German market and based on WpHG-filings, including filings below the 25 percent threshold. Their study was a comparison of announcement effects on stock prices for block purchases by hedge funds and private equity firms and it was later published in an amended version by Mietzner, Schweizer and Tyrell (2011). Significant, positive abnormal announcement returns were found for both hedge fund and private equity transactions. The mean percentage of shares owned after the hedge fund transactions was 12.2 while for private equity transactions the mean value was 52.4. This illustrates how private equity firms tend to acquire majority stakes in contrast to hedge fund firms. The latter investors prefer taking minority stakes.

A block of 50 percent of voting rights is considered as a majority stake because it enables the blockholder to control the general course of business (see Jenkinson and Ljungqvist (2001) for a detailed explanation of different block sizes and the corresponding control rights in Germany).

Table I.2. Research Methodologies and Results of Studies on Shareholder Activism; All three studies focus on shareholder activism by hedge funds in Germany

	Bessler et al. (2010) (Unpublished working paper)	Drerup (2010) (Unpublished working paper)	Stadler (2010) (Doctoral thesis)
Observatio period	2000-2006	1999-2010	1995-2008
Definition of type of activism and findings	Define hedge funds as *"aggressive"* or *"passive"* based on reputation; 85 of 231 events classified *"aggressive"* but observe only 19 actual cases of *"interventions"* where the investor puts public pressure on the target firm	Defines hedge fund as *"active"* or *"passive"* based on reputation; 142 events classified as *"active"*; no further investigation of objective of investor	Defines an event (as opposed to defining the hedge fund) as *"active"* if the investor announces activist intentions; 37 of 136 events *"active"*, thereof 18 events aimed at replacement of supervisory board or management board
Short-term stock price effect (cumulative abnormal return (CAR))	Highly significant CAR of approximately 3-4.5% across all events	Highly significant CAR of approx. 3-4% for active events; no significance for passive events	Highly significant CAR of approx. 3-4% across all events, CAR higher for active events
Long-term stock price effect (buy-and-hold abnormal return (BHAR), calendar time portfolios (CTP))	Up to 3 years BHAR analysis: results positive in the magnitude of 11-24% and highly significant; up to 3 years CTP analysis: results positive but below 1.5% in magnitude, weaker significance levels, abnormal return seems to disappear when time horizon increases	1 year BHAR analysis: results negative but insignificant; CTP using different regression estimation techniques: mixed results, author concludes neither creation nor destruction of value	Up to 6 months BHAR analysis: mixed results, mostly insignificant
Changes in operational performance	No analysis	Concludes no significant effect in measures such as return on equity, return on assets, free cash flow / sales and leverage ratio in comparison to control firms	No analysis

Bessler, Drobetz, and Holler ((2008) and (2010)) and Drerup (2010) in their working papers focus exclusively on shareholder activism in Germany. Their studies cover a specific type of minority shareholder, hedge funds. Stadler (2010) has published a doctoral thesis on the same topic. Table I.2. summarizes the research methodologies and findings of these three studies. The purpose is to summarize results but also to point out difficulties in the research methodologies applied in these studies.

All three studies find similar positive abnormal stock returns in a short-term event study. When considering long-term stock price effects, the results do not match. The authors find positive effects, no effects as well as negative effects. The outcome apparently depends on the event study time horizon and the methodology applied. Detecting long-term abnormal stock returns can prove to be difficult for several reasons, as pointed out by Fama (1998). Even though contemporary research methodologies control for market factors, an event like the bankruptcy of Lehman Brothers Inc. may still bias the results of a long term event study. The bankruptcy of Lehman Brothers, which occurred on 15 September 2008, was the largest bankruptcy in U.S. history. It led inter alia to unprecedented stock market reactions.

There is another possible explanation why not all of the three long-term event studies as presented in Table I.2. display significant positive abnormal returns. The reason could be the German corporate governance framework. German corporations have a supervisory board and large German corporations mostly have co-determined supervisory boards. Co-determination not only takes place on the supervisory board level but also in the form of workers' councils (Betriebsrat). An activist shareholder does not necessarily take any action that affects a corporation's workforce, but if he does, the workforce has the right to join the conversation. Any significant activist action that does not affect the workforce will still be in the area of responsibility of the supervisory board. Therefore, the supervisory board can, from an activist's view, be regarded as the pivotal corporate governance authority. If the activist is unable to obtain supervisory board representation, then he might fail in his attempt to bring about change.

A long-term event study on shareholder activism in Germany should investigate cases of actual supervisory board representation by activist shareholders. Drerup (2010), by not finding any significant improvement in operational performance measures (see Table I.2.), somewhat confirms this hypothesis, since his study is neither based on cases of supervisory board representation nor exclusively focuses on cases of *actual* activism. Instead, Drerup distinguishes between "active blocks" and "passive blocks" based on the reputation of the hedge fund as activist or as non-activist.

A certain number of studies for the German market have investigated related research topics. Achleitner, Betzer and Gider (2010) document empirical evidence that hedge fund and private equity fund activities in Germany are driven by corporate governance improvements. Achleitner, Andres, Betzer, and Weir (2011) find that private equity investments on the German stock market generate positive abnormal stock returns of 5.9 percent for shareholders around the event day. Schäfer and Hertrich (2011), in their working paper, take a different approach to investigate shareholder activism in Germany. Their study presents empirical evidence on the success of *formal* activism. Formal activism includes agenda proposals, counter-motions, and opposition at shareholder voting. The study covers 87 annual meetings of the 30 DAX-listed German corporations between 2008 and 2010. The authors conclude that formal activism in Germany remains a relatively ineffective instrument for shareholder engagement and that voting at annual general meetings remains an "approval process for current management and the supervisory board" rather than a forum for activism.

Fleischer (2008) has published a study of the German capital market's legal framework for financial investors. He points to the lack of research on shareholder activism in Germany, alluding the comprehensive analysis of Brav et al. (2008) for the U.S. market. Fleischer remarks that an empirical investigation for the German market of similar format is not available. Kaserer, Achleitner, von Einem, and Schiereck (2007) have published a comprehensive study for the German market covering private equity activities. Private equity firms, however, usually invest in majority stakes or acquire whole firms. A majority stake or full control – as opposed to a minority stake – generally allows the investor to "dictate" the course of business given the voting power tied to the equity stake. Influencing corporate decision-making through minority stakes is more complex.

Brass (2010) investigates the legal framework for hedge fund activism in Germany. He provides three examples of firms where hedge funds have allegedly bought an activist stake. The three examples are the large, DAX 30-listed corporations TUI, SAP and Adidas. While TUI was actually targeted by various activist shareholders during the past decade, SAP and Adidas were not. There were no signs of activism. There was no regulatory filing of voting rights by any activist. There were no motions brought forward at annual meetings. Brass had not been able to base this part of his study on empirical facts given a lack of empirical research.

Despite the above mentioned studies, the phenomenon of shareholder activism in Germany appears to be poorly researched. Empirical research is at an early stage. A comprehensive study of activism, including different types of activists and different levels of activism, across a sufficiently long observation period and taking into account the peculiarities of the German corporate governance framework is missing at this point. Many effects of shareholder activism, such as the post-investment change to the annual meeting attendance rate of target firms, have not been researched at all.

I.5. Research on Shareholder Activism in the U.S.

For U.S. corporations, more detailed data on company financials, capital markets and corporate governance is available in machine readable form. In addition, there is a substantially larger universe of firms. The availability of data is not restricted to share price information, balance sheet data, income statement positions or items from the cash flow statement. Gompers, Ishii and Metrick (2003) study 24 corporate governance rules across as many as 1,500 U.S. firms dating back to 1990 to come up with conclusions regarding the effect of shareholder rights on share prices. The source for the corporate governance rules is the Investor Responsibility Research Center. There is a certain depth and breadth of data available to researchers analysing the U.S. market. As Drobetz, Schillhofer and Zimmermann (2004) put forward, this kind of detailed data is not readily available for the German market, especially not for small- and medium-sized firms (Kim, Kitsabunnarat-Chatjuthamard and Nofsinger, 2007).

Studies covering different topics in shareholder activism in the U.S. regularly appear in major finance publications (see for example Journal of Finance, Journal of Financial Economics, and Review of Financial Studies). This raises the question if empirical research on shareholder activism in Germany adds any value to the existing literature. The answer is: yes, because the two regulatory frameworks are different.

Fleischer (2008) points out that it is not possible to simply apply the results from U.S. studies to Germany given the differences in the corporate governance structures. Prigge and Steenbock (2002) go even further, formulating that the organisational set-up of U.S. (institutional) investors differs from their German counterparts. Therefore results from U.S. studies cannot be applied to the German market.

Peculiarities of the U.S. corporate governance system include non-binding voting on shareholder proposals (Levit and Malenko, 2011), the Delaware General Corporate Law (Dooley and Goldman, 2001), staggered boards (Bebchuk and Cohen, 2005), anti-takeover devices such as poison pills (Bizjak and Marquette, 1998), and certain tax rules governing investments by institutional investors, for example mutual funds. Hellgardt and Hoger (2011) provide a comprehensive overview of the many differences between U.S and German corporate law.

Research on shareholder activism in the U.S. ranges from studies on the effects of shareholder proposals (Gillan and Starks, 2000), private negotiations on corporate governance issues by institutional investors (Carleton, Nelson and Weisbach, 1998), labour union-sponsored activism (Prevost, Rao and Williams, 2012), and activist block share purchases, providing the investor with partial corporate control (Bethel et al., 1998), to the new shareholder activism by hedge funds (Brav et al., 2008) as well as other private investors (Klein and Zur, 2009). Some institutional investors, when dissatisfied with the management of a firm "vote with their feet", which means they simply sell their stake ((Parrino, Sias and Starks, 2003), (Helwege, Intintoli and Zhang, 2012)). Karpoff (2001) and Gillan and Starks (2007), in their surveys, summarize key research findings and also describe the history of shareholder activism in the U.S. The results of the studies are mixed. Generally speaking, the success of shareholder activism from a shareholder value perspective depends on the ability of the activist to force changes.

I.6. Development of Hypotheses

The first hypothesis is straightforward. I hypothesize that the activity of activist shareholders in Germany has increased during the past two decades.

Furthermore, if activist shareholders are really active shareholders, engaging in the firm's corporate governance, then they should participate and cast their votes at the annual general meeting of that firm. The attendance at the annual meeting following the announcement of an activist stake should therefore rise. The investigation will also include a consideration of the possibilities of influencing the target firm as a minority shareholder given the observed attendance rates.

It is assumed that agency costs exist. If an activist shareholder is able to reduce agency costs then the announcement of an activist stake should lead to positive stock price reactions. The stock price reaction is an anticipated reduction in agency costs resulting in an increased overall market capitalisation of the target firm. This will be complemented by an analysis of the drivers of abnormal stock returns measured around the announcements.

Given the fixed term nature of the German supervisory board's term of office I assume that a minority shareholder who tries to gain partial control of the corporation prefers to invest in companies whose next supervisory board election is closer. The frequency of announcements of activist stakes should therefore increase with the supervisory board election moving closer.

A single activist investor holding a small percentage of a firm' stock can increase his influence and more likely impose his agenda through convincing other potential activists to invest in the same firm. This would also be a reasonable approach to diversify the economic risk from one investor to multiple investors. I assume that simultaneous investments by multiple activists can be observed. This approach is based on the assumption that the investors either hold less than 30 percent of voting rights collectively or that their actions do not constitute an acting-in-concert in the sense of section 35 WpÜG (German takeover code; Wertpapiererwerbs- und Übernahmegesetz; abbreviated WpÜG). Such an acting-in-concert would require the investors to submit a mandatory takeover bid for the whole firm.

The past century's stereotypical view of German finance was that the influence of large shareholders such as founding families and banks had been substantial (Jenkinson and Ljungqvist, 2001), market transparency was low (Becht, 1997), small investor's virtually did not participate in corporate governance (Shleifer and Vishny, 1997) and minority shareholders' rights were weak ((Rafael La Porta, Lopez-de-Silanes, Shleifer and Vishny, 1998), (Rafael La Porta, Lopez-de-Silanes, Shleifer and Vishny, 2000a). I assume that because the influence of banks has decreased in the past two decades ((Vitols, 2005), (Dittmann et al., 2010)) and regulatory initiatives have further increased accountability and transparency in corporate Germany ((Nowak, 2004), (Goergen et al., 2008)) prior research findings need to be reviewed. There is no formal test for finding weak or strong shareholder rights. However, observing an increased number of cases of minority investments from international investors, and possibly also supervisory board representation, can be viewed as evidence to, at least, question the stereotypical view of German finance. This especially includes the weak shareholder rights view.

I.7. Remarks on the Collection of Data

In 1995, a new mandatory disclosure standard was introduced in Germany based on the European Union's Large Holdings Directive (88/627/EEC). This new standard was named Wertpapierhandelsgesetz (German securities trading act). It henceforth required holders of voting blocks of 5 percent or above to file a disclosure with the Bundesaufsichtsamt für den Wertpapierhandel (BAWe, German Federal Securities Supervisory Authority). The BAWe later changed its name and is now the Bundesanstalt für Finanzdienstleistungsaufsicht (BaFin, Federal Financial Supervisory Authority). Before the enactment of the WpHG in 1995, the threshold for disclosure was at 25 percent.

Table I.3. Possible Definitions of Activist Event and Evaluation for Event Study Purposes

Activist Event	Event Date	Evaluation for Event Study Purposes
Research Methodology Applied		
WpHG-disclosure at 3%- and 5%-threshold	Date of disclosure of stake by the corporation	Does not contain any information on intentions and purpose of investment by the shareholder; further research on investor's intention necessary
Media coverage	Date of issue of press article	Good visibility in terms of capital markets impact; press articles describe activist intention very well in most cases; includes equity stakes below thresholds for regulatory disclosure; possible bias towards larger companies given higher frequency of media coverage needs to be controlled for
Shareholder gaining supervisory board representation	Date of disclosure (of the WpHG-filing) by the corporation	Clearly identifies shareholder's approach as an activist; no bias towards larger companies; using date of WpHG-filing as event date possible; probably limited number of observations
Other Possible Research Methodologies		
Self-definition of investor as activist and filtering all WpHG-filings for disclosures from these "known" activists	Date of disclosure of stake by the corporation	Less powerful differentiation between activist and non-activist events (does not substitute for researching every single event)
Defining, who is "known" to be an activist shareholder, and filtering all WpHG-filings for disclosures from these "known" activists	Date of disclosure of stake by the corporation	Less powerful differentiation between activist and non-activist events (does not substitute for researching every single event)

Shareholder proposals or counter-motions at annual general meeting ("formal" activism)	Date of disclosure of stake by the corporation or date of annual general meeting	Shareholder proposals and counter-motions are often filed by retail shareholder associations, which means, there is only a limited possibility to control for the size of the "activist" stake; limited visibility both in terms of media coverage and capital markets impact; impact of single proposal cannot be analysed in isolation given that at each annual meeting several proposals will be voted upon within a few minutes / hours
WpHG-disclosure at 10%-threshold	Date of disclosure of stake by the corporation	Contains detailed information on intention and purpose of investment by the shareholder; respective legislation was only introduced in 2007; only a fraction of activists cross the 10%-threshold; therefore observation period and number of observations very limited

Therefore the identity of minority shareholders holding less than 25 percent of voting rights as well as the size of the equity stake was filed with an authority and stored in a central database. Instead, it had to be hand-collected from sources such as voluntary disclosures in annual reports and privately-run data collections. Needless to say, an observation period starting after 1995 will yield a larger number of observations. This is relevant, since the size of the German capital market in terms of number of firms is lower compared to the number of U.S. publicly-listed firms.

From January 1999 onwards, the number of activist events in Germany slowly started to increase (Stadler, 2010). The chosen observation period therefore starts in January 1999. It reaches until May 2011. Stadler's (2010) data sample includes events before the year 2008. It focuses on the identity of the investor as a hedge fund and excludes certain industries such as financial services. An empirical analysis, that has a longer observation period, includes non-hedge fund events in the sample and includes all industries will yield a larger sample. It also offers more possibilities for differentiated analysis, for example between activist events and potentially activist events. If needed, investor or target firm characteristics can still be controlled for.

I introduce four *Activity Levels* for categorising each event according to the activist's level of engagement and the capital market's perception. The four Activity Levels will be described in more detail in the first study. This approach results in obtaining a comprehensive sample of events with different levels of shareholder activism.

Table I.3. illustrates the possibilities for identifying potential shareholder activist events and the corresponding research methodologies. The research methodology applied allows for an analysis that yields the most comprehensive and powerful results.

The research methodology applied also represents a quite work-intensive approach. It takes several steps of manually reviewing databases and events.

Two external events that led to heavy stock market reactions and an overall increase in stock-market volatility fall into the observation period. These events were (i) the terrorist attacks on the World Trade Centre in New York City on 11 September 2001 and (ii) the bankruptcy of global banking institution Lehman Brothers Corporation on 15 September 2008. The simultaneous takeover of Merrill Lynch Corporation by Bank of America on the same day as Lehman Brothers insolvency adds to the series of events. In the 30 calendar days following 11 September 2001, there are no events. Following 15 September 2008, within 30 days there were four announcements regarding investments by potential activist shareholders. This will be taken into account with respect to the analysis. Figure I.3. shows the quarterly distribution of events across the final sample.

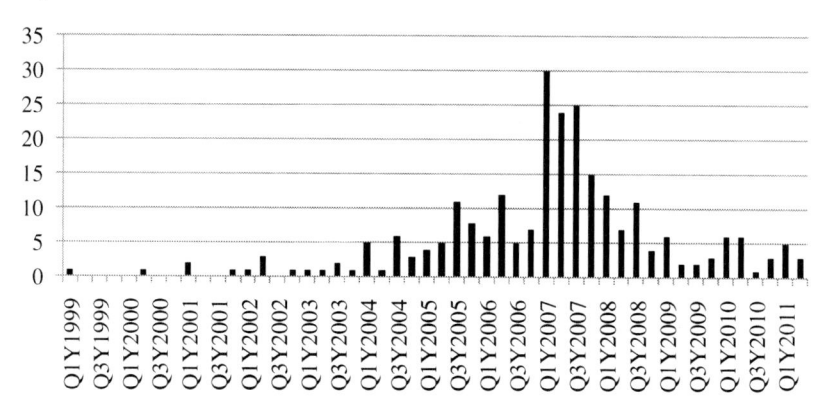

Figure I.3. Quarterly Distribution of Events

The figure shows the final sample of 253 events. The vertical axis shows the number of events per quarter. The process of data collection is brought forward in detail in the first study.

I.8. Presentation of Studies and Contribution to Literature

I.8.1. First study: Shareholder Activism in Germany – An Empirical Review

"Shareholder Activism in Germany – An Empirical Review" is the first study to analyse most of the available information, both qualitative and quantitative, on the activity of activist shareholders and potential activist shareholders in Germany in the past ten to fifteen years.

The contribution to literature of the first study is (i) providing a comprehensive overview of the literature that is dedicated to shareholder activism in Germany, (ii) providing a detailed description on the process of corporate governance variables data collection, especially for small- and medium-sized German firms, (iii) introducing Activity Levels for the purpose of categorising activist events, (iv) an analysis of the timing of the activist investments, (v) an analysis of the changes to annual meeting attendance rates following activist investments, (vi) a theoretical consideration on the possible effects of shareholder activism with respect to the measured attendance rates. Further analyses such as a major corporate events overview complement the existing literature.

The test statistic for the difference-in-means test applied to detecting significant changes in annual meeting attendance rates, before and after the disclosure of an activist stake, is the standard test statistic

$$t = \frac{\bar{X}_1 - \bar{X}_2}{\sigma_{\bar{X}_1 - \bar{X}_2}}$$

where $\bar{X}_1$ and $\bar{X}_2$ are the means of the two samples, and $\sigma_{\bar{X}_1 - \bar{X}_2}$ is a measure of the variability of the differences between the sample means. This methodology is also applied to detecting significant differences between the sub-samples in the NewBET analysis in the second study. The first study is complemented by the more quantitative approach of the second study.

I.8.2. Second Study: Activist Shareholders, Abnormal Returns, and the German Aufsichtsrat

"Activist Shareholders, Abnormal Returns, and the German Aufsichtsrat" is an event study applying state-of-the-art event study methodology and taking into account the peculiarities of the German corporate governance system. The contribution to literature of this study is (i) an explanation of the legal rules governing the German supervisory board with a special focus on these rules' implications for conducting an event study analysis, (ii) an event study on four different activity levels, (iii) a comparison of the event study results with respect to investments by *non*-activist Anglo-American equity investors, (iv) an analysis of the event study results considering the timing of the investment ,and (v) an analysis of the drivers of abnormal returns through ordinary least squares regression that incorporates a set of explanatory corporate governance variables.

The analysis of the timing also has implications for researchers. It provides a tool, that is an additional control variable with virtually no correlation to frequently used control variables. This can improve the results of regression analyses (Auer, 2007).

The results of the second study are also meaningful for practitioners such as equity fund managers. Abnormal returns are largely driven by the credibility of the activist effort. An important factor in terms of credibility is apparently the remaining term of office of the incumbent supervisory board members. More credible events result in higher post-event abnormal returns.

Abnormal returns to the stock price of firms targeted by activist shareholders are calculated using the approach introduced by Brown and Warner (1985). The market model is applied to calculate the abnormal return ε_{it} for each target firm i on the event day t:

$$\varepsilon_{it} = R_{it} - (\hat{\alpha}_i + \hat{\beta}_i R_{mt}).$$

The difference between the actual return R_{it} on the event day t and the expected return on day t, as calculated by the market model, is the abnormal return, where R_{mt} is the return on the market portfolio on day t. α_i and β_i are are coefficients estimated from an ordinary least squares-regression of the daily stock returns of target firm i on the market return over an estimation period. I choose a long estimation period of 252 trading days before the event date t $[t-282; t-30]$ to obtain unbiased estimations of the coefficients α_i and β_i. My approach follows Achleitner, Andres, Betzer, and Weir (2011) and Mietzner, Schweizer, and Tyrell (2011) in using the broad, value-weighted German C-DAX index as a proxy for the market portfolio.

Daily abnormal returns can be calculated as the simple average of the cross-section of abnormal returns in the sample. The cumulative abnormal return (CAR) for event periods longer than just the event day itself is calculated by building the sum of daily abnormal returns. I indicate CARs for event periods of up to 40 days surrounding the event $[t-20; t+20]$. For a further explanation of the event study methodology applied see MacKinlay (1997) and Corrado (2011).

Statistical significance of CARs is tested using the standard t-test (Brown and Warner, 1980). The Boehmer, Musumeci, and Poulsen (1991) T-test is applied as a robustness check. The Boehmer T-test controls for event induced variance. It produces more appropriate rejection rates when the null hypothesis is true and equally powerful tests, when the null hypothesis is false. The Wilcoxon signed rank-test (Wilcoxon, 1945), a non-parametric test statistic, is applied to test for significance of median CARs.

The drivers of CARs are analysed through multivariate ordinary least squares regression of the form

$$y = b_0 + b_1 x_1 + b_2 x_2 + ... + b_i x_i$$

where b_0 is a constant, b_i the regression coefficients, x_i is the respective control variable and y as the dependent variable is the cumulative abnormal return for each observation. The regression will be estimated using heteroskedasticity-robust standard errors (White, 1980).

Table I.4. shows the matrix of Pearson sample correlation coefficients for the control variables used in the regression models. There is virtually no correlation in any of the control variables. The only exception is the variable *Cofull* (full co-determination) and *Loginitialstake* (logarithm of the initial percentage stake acquired by the activist) with a correlation of -0.31. This correlation is quite intuitive, since in larger firms activists can only acquire smaller stakes. And larger firms usually have co-determined boards. By definition, there is a negative correlation between the variables *Cothird* and *Cofull*. The existence of correlation can impair the quality and reliability of regression results.

Table I.4. *Matrix of Pearson Sample Correlation Coefficients; CAR01 is the [0; +1]-event window cumulative abnormal return. CAR01 is the dependent variable of the regression analyses. All control variables are explained in Table III.21. (Results of Ordinary Least Squares Regression). Perf. stands for performance*

Activity Level (3 & 4)	Log initial stake	Wolfpack	NewBET2	Prior 12m perf.	RoE	Cash to Assets	AGM attendance rate	Cothird	Cofull	CAR01	
1.000	-0.019	0.102	0.071	0.079	0.031	-0.002	-0.091	-0.010	0.126	0.219	Activity Level (3 & 4)
	1.000	-0.019	-0.084	0.079	-0.040	0.042	-0.081	-0.018	-0.311	0.147	Loginitialstake
		1.000	-0.089	-0.146	0.016	0.002	0.052	-0.056	0.075	-0.080	Wolfpack
			1.000	0.126	-0.082	-0.007	-0.019	-0.010	-0.066	0.132	NewBET2
				1.000	0.094	0.050	-0.080	0.092	-0.068	-0.021	Prior 12m perf.
					1.000	0.044	0.039	0.102	-0.085	0.030	RoE
						1.000	-0.006	0.026	-0.148	-0.095	CashtoAssets
							1.000	-0.055	0.046	-0.177	AGM attendance rate
								1.000	-0.326	0.024	Cothird
									1.000	-0.020	Cofull
										1.000	CAR01

I.8.3. Third Study: Weak Shareholder Rights? – A Case Study of Cevian Capital's Investment in Demag Cranes AG

"Weak Shareholder Rights? – A Case Study of Cevian Capital's Investment in Demag Cranes AG" is an enhanced case study of activist investor Cevian Capital's investment in Demag Cranes AG in 2010. It is enhanced in the sense, that differences and similarities between the German stock corporation and the U.S. corporation are discussed. The study complements this dissertation as it is dedicated to a single incident of shareholder activism going further into detail. It is beneficial for the understanding of German corporate governance and therefore the first and the second study.

The contribution to literature is (i) identifying shortfalls of prior international research on German corporate governance, (ii) creating a better understanding of the differences between the U.S. corporation and the German stock corporation, (iii) questioning the weak shareholder rights view of the past century promoted by La Porta, Lopez-de-Silanes, Shleifer and Vishny ((1997) and (1998)), (iv) providing a table of activist shareholder supervisory board representation for the first time and (v) presenting a case of successful and *consensus-oriented* activism. The empirical evidence presented, documents how small shareholders actively participate in German corporate governance and benefit from their active participation.

II. Shareholder Activism in Germany – An Empirical Review

Abstract

According to agency theory shareholders can reduce agency costs and increase firm value through monitoring the management. The international literature claims that shareholders of German corporations are weak owners given the peculiarities of the corporate governance system and that their activities are insignificant. This study examines the monitoring activities of activist minority shareholders in Germany between January 1999 and May 2011. A sample of 253 cases of potentially activist stakes in 140 target firms is analysed. Open disputes between shareholder and activist occur in approximately one out of four cases. Some disputes lead to a dismissal of members of the supervisory board or the management board. The average activist stake measures 5.9 percent and therefore activist investors frequently leverage their influence by simultaneously targeting firms.

II.1. Introduction

"Shareholder activism has not played an important role in Germany so far." This is a quote from the introduction to the doctoral thesis of Max Steiger (2000) on the impact of institutional investors on corporate governance. It describes very well the situation in Germany until the year 2000. In 2004, a hedge fund publicly criticised the management of stock market-listed soccer club Borussia Dortmund. In 2005, activist shareholders removed the management of Deutsche Boerse. Prior to the 2007 annual meeting of CeWe Color Holding, there was a public dispute about future strategy between the firm's management and a group of minority shareholders. Shareholder activism in Germany is obviously becoming a more prominent topic.

Gillan and Starks (2007) define activist shareholders as "investors who, dissatisfied with some aspect of a company's management or operations, try to bring about change within the company without a change in control". This study goes even further and applies the concept of *potentially* activist shareholder because shareholder activism is not always visible (Becht, Franks, Mayer and Rossi, 2010b). Furthermore, a number of studies classify activist shareholders according to their historic track record and their reputation (see for example Bethel, Liebeskind and Opler (1998)); in Germany, however, most investors do not have a track record. In addition, there is no equivalent to the U.S. SEC 13D filing, that requires investors holding 5 percent or more of voting rights to disclose their intentions to the public (the threshold for such disclosure is at 10 percent and it was only introduced in 2007). Studies on shareholder activism on the U.S.-stock market in many cases present no formal definition, but point to the investor's regulatory SEC 13D filing (see for example Greenwood and Schor (2009)).

There is no formal or legal definition of the concept of *activist shareholder*. Rather, the German stock corporation act stipulates a certain set of shareholder rights. Section 112 Investmentgesetz (German investment company act) provides a definition of "hedge fund". However, hedge funds are only one certain type of activist shareholder (Bassen, Königs and Schiereck, 2008). Moreover, not all hedge funds invest in public equity and not all hedge funds that invest in public equity are activists. Shareholder associations (Aktionaersvereinigungen) are excluded from the analysis as well as the group of professional plaintiffs (Berufsklaeger). The professional plaintiffs' approach is, to pressure corporations through appealing shareholder resolutions. Their focus is clearly not on creating shareholder value for all shareholders through monitoring and reducing agency costs (for a discussion see Baums, Drinhausen und Keinath (2011) and Schiereck and Thamm (2012)).

The modern corporation is characterised by the separation of ownership and control. This separation can lead to an inefficient use of company resources by the management that is ultimately in control (Berle and Means, 1932). The Principal-Agent-Theory is the theoretical framework frequently applied to discuss and solve the conflicts of interest that arise from this separation ((Jensen and Meckling, 1976), (Fama and Jensen, 1983)). Through efficient monitoring of the management agency costs can be reduced. If activist shareholders are better monitors than other more passive shareholders, this can have a direct effect leading to an increase in firm value.

The purpose of this study is to gather as information as possible on target firms, activist strategy and effects of activism in Germany. While recent studies on shareholder activism concentrated on hedge fund investors, and therefore the identity of the activist, particular attention will in this study be paid to the activism itself. Several groups of potentially activist minority shareholders will be included in the analysis.

The analysis proceeds as follows. The next paragraph presents a literature review and formulates hypotheses. The process of data collection, which was quite lengthy, is described in the section that follows. The results section first presents some descriptive statistics and then analyses the activists' investment approach and its effects. The last paragraph is a summary of results.

II.2. Literature Overview

Empirical investigations on shareholder activism in Germany include Achleitner, Betzer and Gider (2011), Bessler, Drobetz and Holler (2010), Drerup (2010), Mietzner, Schweizer and Tyrell (2011) as well as Stadler (2010). The contributions by Mietzner et al. (2011) and Achleitner et al. (2010) compare hedge fund and private equity activities, while the other three studies focus on hedge fund activism. Table II.5. provides an overview of how each team of authors collected their data. The number of observations in each study depends on the definition of "activist shareholder", the observation period and the databases included in the analysis. The frequency of known cases of shareholder activism, including hedge fund activism, before the year 2000 was low (Stadler, 2010).

Jenkinson and Ljungqvist (2001) presented an early study of shareholder activism ("hostile stakes") in Germany. The authors concluded that the German system of corporate control and governance was more active than had previously been suggested. The number of observations is 17 between 1987 and 1994. Croci (2007) in his study investigated the influence of active investors ("corporate raiders") in Europe between 1990 and 2001. His results showed superior stock-picking ability by these raiders, but not necessary corporate governance qualities. The sample used by Croci included 47 observations from the

United Kingdom, 25 observations from Italy, 24 observations from France, 21 from Switzerland and only 13 observations from Germany. The structure of this sample underscores the remote impact of shareholder activism before the turn of the millennium. Needless to say, the quality and quantity of available information on shareholder's activities has increased substantially in the meantime (Croci, 2007). In the past twelve years shareholder activism has become a more prominent topic. Today, empirical studies put their focus on activist shareholders given the improved availability of information.

This study adds to the existing empirical literature. Some basic questions, that have not been answered so far, will be discussed. The process of data collection aims at gathering a comprehensive sample to answer open research issues. Therefore it includes all types of potentially activist shareholders, and not just for example hedge funds. The study distinguishes between activist and potentially activist investments. This provides the possibility for additional analyses.

The contributions by Prigge and Steenbock (2002) and Fleischer (2008) show how this research topic is just emerging. Prigge and Steenbock's paper is a theoretical study on pension fund activism and evaluates corporate governance opportunities and threats from changing the current retirement benefits regime in Germany. Conducting an empirical investigation on this topic would have proved to be difficult, given the lack of data. In part, Prigge and Steenbock's findings are based on studies of pension fund activism in the U.S. Fleischer (2008) directly mentions to the lack of empirical studies on shareholder activism in Germany:

"Various studies of shareholder activism on the U.S. market conclude that (…). Such comprehensive studies for the German market are still missing." (quote translated)

Fleischer further remarks, that studies on private equity transactions do exist. Private equity investors, however, mostly take majority stakes and only have a limited interest in acquiring minority stakes in the target firm. There is apparently a research gap with respect to activist minority stakes in Germany. Fleischer adds:

„*Now, given the differences between the German and the U.S. corporate governance framework, one cannot easily apply the results of U.S. studies to the German market.*" (quote translated)

This is an important detail. Many differences exist between the German and the U.S. corporate governance system. Some of these differences are in fact substantial. Hellgardt and Hoger (2011) in their comparative study on shareholder rights in Germany and the U.S. provide numerous examples. There are two key takeaways. Minority shareholder activism in Germany is poorly researched. Furthermore, the results of U.S. studies cannot simply be transferred to the German market. Instead, emerging research should question every detail before adopting both, research design, and results.

The potential for influencing firm policies as a minority shareholder to some extent depends on the annual meeting attendance rate. A low attendance rate means that minority stakes will have a stronger impact on voting outcomes. Attendance rates at annual meetings of target firms will be reviewed as part of the analysis.

Another purpose of this study is, to find out what types of investors, besides hedge funds, take potentially activist minority stakes. Activist shareholders seem to apply certain investment strategies. For example, Stadler (2010) remarks that activists quite often simultaneously target firms. A question that has not been answered so far is how the timing of the next supervisory board election influences investment decisions. This is of relevance since the supervisory board is a pivotal corporate governance authority in Germany. Supervisory board members are elected for a fixed term of up to five years. Members of the board of directors in the U.S. are (re-)appointed each year in the case of an unstaggered board of directors.

The investigation into the effects of shareholder activism adds to the findings of Stadler (2010) who studied hedge fund activism in Germany. Stadler analyses the demands set forth throughout 37 of the 136 events in his sample. In the case of these events the hedge funds especially ask for changes on the supervisory board and a sale of parts of the company. Whether or not the demands are fulfilled and what major corporate events actually occur is not analysed.

A set of major corporate events is defined. Major corporate events are inter alia changes on the supervisory board, a takeover of the target firm or the initiation of a share buy-back programme. In comparison to Stadler, this study has a longer observation period and includes non-hedge fund activist shareholders. The result is an increased sample size (see Table II.5.).

Table II.5. *Overview of Empirical Studies on Shareholder Activism in Germany; Number of observations refers to the final sample. Private equity fund investments are usually majority investments or complete takeovers. By definition, these investments are not minority investments and therefore not shown in the number of observations column*

Author(s)	Purpose of study	Data source	Period	Number of observations
Achleitner et al. (2010)	Motives of hedge fund and private equity fund investments	-BaFin-database (lists regulatory filings according to the German securities trading act (WpHG)) -Factiva, Lexis Nexis, Google, and investor magazines for classification of investors as hedge funds	1998-2007	96 hedge fund investments
Bessler et al. (2010)	Discussion of hedge fund activism	-Lexis Nexis search for results containing "hedge fund" and name of target firms -Agentur für Unternehmensdaten for WpHG-filings	2000-2006	231, thereof 19 observations where the investor pressures the target firm's management
Drerup (2010)	Discussion of hedge fund activity	-BaFin-database -in addition, database of WpHG-filings by the Deutsche Gesellschaft für Ad-hoc-Publizität, general search of target firm websites, Google and Bloomberg News	1999-2010	278, thereof 142 "active" and 136 "passive" blocks of voting rights
Mietzner et al. (2011)	Analysis of intra-industry effects of hedge fund and private equity investments	-using Eureka Hedge and Credit Suisse Tremont databases to create a list of hedge funds -BaFin-database for regulatory filings by funds from list	2001-2009	72 hedge fund investments
Stadler (2010)	Discussion of hedge fund activism	-combined search for hedge fund names and regulatory filings using several databases (BaFin, Factiva, Genios, Eureka Hedge, others)	1995-2008	136 hedge fund investments (excluding investments in financial services firms)
This study	Study of shareholder activism	-BaFin-database, Bloomberg News, Boersen-Zeitung Pro, dgap.de, Lexis Nexis, WAI Wirtschaftsanalysen und -informations GmbH (www.hv-info.de), company websites and financial reports	1999-2011	253, thereof 61 observations where investor publicly pressures the target firm or obtains superv. board seat

II.3. Data Collection

There is no central database storing cases of shareholder activism. The sample was gathered manually on an observation-by-observation basis.

Becht, Franks and Grant (2010a) as well as Gospel, Haves, Vitols, Voss and Wilke (2009) in their studies of shareholder activism in Europe provide a list of activist investors. This list is completed with information from the studies mentioned in Table II.5. and using the database Lexis Nexis. Drawing on U.S. studies is not necessarily helpful. Many activists that have a solid track record in the U.S. do not invest in German corporations. One prominent example is the "corporate raider" Carl Icahn, who has not reported any direct investments in German corporations in the past so far.

The fund managers' personal names are collected in addition to the name of the activist fund or investment vehicle. Some fund managers build positions in target firms through different legal entities even though the decision-taking person is the same. This approach was inter alia applied by hedge fund manager Florian Homm. He invested in German corporations through various entities (Absolute Capital Management, Absolute Octane Fund, Absolute Return Europe Fund, CSI Asset Management Establishment, European Catalyst Fund, FM Fund Management and Fortune Management). These entities were mostly incorporated in off-shore jurisdictions such as the Cayman Islands.

An investor is deemed a potential activist if he (i) has received public attention for past activist investments (ii) announces an activist attention for the first time, (iii) is labelled activist in the financial press. An investor whose fund (iv) has sufficient capital resources and presumably a performance-based compensation scheme in place is labelled as activist, if the fund can be regarded as independent from third-party interests. This last criterion is not fulfilled by many institutional investors, for example by some of the largest equity funds. Institutional investors often invest in the same corporations as their competitors and therefore remain passive investors. Business relations between the investor and the portfolio companies are not infrequent. These business relationships can lead to conflicts of interest (Kahan and Rock, 2007). Certain industry trends, such as benchmarking and indexing, lead to an increase in shareholder passivity (Schmolke, 2007).

Some investors, especially hedge funds, provide a self-classification as activist or non-activist. This self-classification is not used a primary source for building a list of potentially activist shareholders, since the investors' self-classification and the observed approach are not necessarily consistent.

The next step is a Lexis Nexis search using the list of potentially activist shareholders. The purpose of the search is to now find cases of potentially activist investments and determine the target firm and the date (year) of the investment. By using this approach, stakes below regulatory thresholds can be identified. The regulatory threshold according to the German securities trading act WpHG was at 5 percent before 2007 and it is now at 3 percent.

It is important to trace the first date when a stake was disclosed to the public - either through a regulatory filing or through media coverage. This information is for example not included in individual WpHG-filings. The relevant date, however, is the date when the capital market first hears about the investment. Subsequent disclosures of regulatory filings by the same investor and with respect to the same target firm occur quite often. These filings can lead to wrong conclusions about the actual event date.

Table II.6. Comparison of Disclosure Requirements in the U.S. and in Germany; Displayed are the regulatory disclosure requirements according to WpHG and U.S. Securities and Exchange Commission rules. In Germany, disclosure requirements also apply to stakes of 15%, 20%, 30%, 50% and 75%. Non-compliance with disclosure rules can lead to a penalty and temporary loss of voting rights according to sub-section 28 WpHG. This does not affect the shareholder's claim on company profits

Statute	Notification threshold	Disclosure requirements	Requirements before 2007
Section 21 sub-section 1 WpHG	3%	Name of beneficial owner, name of issuer (target firm), size of stake (voting rights), date of transaction	Notification threshold at 5%
Section 21 sub-section 1 WpHG	5%	Same as above	Unchanged
Section 21 1 / section 27a sub-section 1 WpHG	10%	Same as above, in addition: (i) purpose of investment, (ii) any intentions to increase stake within the next 12 months, (iii) any intentions to influence the target firm's corporate governance, (iv) any intentions to propose changes to dividend policy	No additional disclosure requirements
U.S. Securities and Exchange Commission Rule 13D	5%	Name of beneficial owner, name of issuer (target firm), size of stake (voting rights), date of transaction, financing of transaction, purpose of investment, disclosure of any intentions related to changes to management, sale of target firm or liquidation of target firm	Unchanged

Table II.6. shows legal disclosure requirements in Germany according to WpHG including recent changes to the law. The fourth and last row of the table describes disclosure requirements in the U.S.

Table II.6. illustrates two important issues. As mentioned before, U.S. studies such as Brav, Jiang, Partnoy and Thomas (2008) or Greenwood and Schor (2009) can largely rely on filings with the U.S. Securities and Exchange Commission, when categorising equity ownership as activist or non-activist. In Germany, transparency has increased with the lowering of the disclosure threshold to voting blocks of 3 percent. However, the investor's *intentions* do not have to be disclosed for voting share blocks below the 10 percent threshold. As will be shown later, only few potential activists acquire stakes of 10 percent or above. Moreover, the articles of incorporation (Satzung) can theoretically provide that the corresponding rule (sub-section 27a WpHG) is not applied and intentions, therefore, do not have to be disclosed.

The process of data collection leads to the identification of several hundred observations. In the next step, I look for the exact date of the disclosure and the size of the disclosed stake. The date of the disclosure, which will later be the event date, is the date when the capital market first hears about the potentially activist investment. In any case where I find two different dates, the earlier date is chosen. The size of the stake is measured as the percentage of voting rights as stated in the disclosure or the corresponding news article. Later increases in the stake will be taken separately. Stadler (2010) remarks that multiple disclosures regarding the very same investment are not an exception but the rule. These multiple disclosures are partly due to corrections of mistakes in earlier disclosures. Anecdotal evidence points to attempts of obscuring the exact date of investment.

The BaFin-database of regulatory filings is reviewed in the next step. Unfortunately, the BaFin-database only lists current holdings. Any notification where the size of the stake falls below the lowest notification threshold will be deleted from the database. In this sense the BaFin-database is not a historic database. With the help of WAI Wirtschaftsanalysen und -informations GmbH and additional sources historical copies of the BaFin-database can be traced back to January 1999.

Since the BaFin-database only lists current holdings, a disclosure date as stated in the database is not necessarily the date of the first disclosure. The date may, in fact, be wrong in many cases, for example because of multiple filings. Therefore, every date needs to be double-checked. I apply another double-check of dates and stakes. I review the actual disclosures documents in the authorised financial newspapers (Boersenpflichtblaetter) Boersen-Zeitung, Financial Times Deutschland, Frankfurter Allgemeine Zeitung, Handelsblatt and Sueddeutsche

Zeitung because the BaFin-database itself does not provide disclosure documents. For clarification, there are three dates attached to one disclosure document. The three dates are: (i) the date when the investor reaches the voting rights threshold (ii) the date when the investor notifies the target firm and the BaFin and (iii) the date when the notification is published by the target firm. Date (i) will be used when analysing the timing of the investment and date (iii) will be used for event study purposes. The rules for disclosure stipulate that the maximum number of trading days between date (i) and date (iii) is nine. Weber and Zimmermann (2011) have published an in-depth analysis of this mechanism. Before 2007, the maximum number of trading days was 18.

Electronic filings of disclosures have become more common in the past five years. The publishing of a disclosure on the internet substitutes for the disclosure in an authorised financial newspaper. This new regulation was introduced in 2007. Several firms, such as the Deutsche Gesellschaft für Ad-hoc-Publizitaet (www.dgap.de) store these electronic filings and facilitate online searches for certain filings including keyword search. A third double-check can be implemented using company websites and annual reports. These sources usually contain information on regulatory disclosures.

If there is any sign for a takeover or a planned takeover of the target firm by the activist shareholder himself, the observation is deleted from the preliminary sample. This is a feasible method to distinguish between minority investments and for example private equity investments leading to a buyout of the target firm – even though this applies to a very small number of observations. In a few cases, private equity firms acquire minority stakes. The Blackstone Group acquired a small stake in Deutsche Telekom and Triton Partners took a position in Curanum. These observations form a part of the final sample.

The final sample consists of 253 investments in 140 target firms between January 1999 and May 2011. It probably accounts for the largest portion of all potentially activist investments in Germany during that period that fulfil the criteria introduced above. The following observations have been excluded from the raw sample: (i) a takeover bid for the target firm was launched prior to the activist's investment (31 observations), (ii) debt-to-equity swap investments that mostly occurred in the event of financial distress of the target firm (20 observations), (iii) investments shortly after the initial public-offering of the target firm (30 observations).

The size of the final sample is consistent with sample sizes of prior studies (see Table II.5.). Drerup (2010) finds 278 observations between 1999 and 2010, Bessler et al. (2010) use a sample of 231 observations between 2000 and 2006. These authors have dedicated their research to hedge fund investments. This study's focus is on potentially activist investments. This does not include non-activist hedge fund stakes and the result is a different sample size.

The final sample is divided into sub-samples according to investor activity level. This allows for a more differentiated analysis. Four levels are introduced. These levels are described in Table II.7.

Table II.7. Introduction of Activity Levels; The total number of observations is 253. Passive investments contains all potentially activist investments but where there is actually no sign of activism

Activity Level	Definition	Investment approach	Number of observations
1	Regulatory filing by new shareholder. No additional information.	Passive investments	108
2	(i) Regulatory filing and news coverage. (ii) For filings below regulatory thresholds: news coverage. There is no visible sign of activism, such as public criticism, on Activity Level 2.		84
3	The investor publicly puts pressure on the target firm's management. This can for example be related to certain strategic decisions of the management board or the supervisory board. This criticism is in most cases published in newspaper articles.	Active investments	41
4	There is a change on the supervisory board and, in some cases, the management board. This change is related to the activist's investment in the target firm.		20

II.4. Descriptive Statistics: Target Firm, Stake Size and Activist Shareholders

Between 1999 and 2003 the annual number of observations is five or lower. The frequency of events starts to increase in 2004 and peaks in 2007 with 94 observations. In 2008 the annual frequency drops but remains on a level above the first years in 2010 and 2011 (see Figure I.3.).

Table II.8. shows the distribution of the sample. It shows the initial stake size and the maximum stake size. All initial stakes are below the 30 percent threshold. Any holding of 30 percent of voting rights or above would trigger a mandatory takeover offer by the investor according to German takeover law (section 35 and section 29 subsection 2 Wertpapiererwerbs- und Übernahmegesetz). The largest stake in the overall sample is 30.33 percent, which is an investment in Ehlebracht AG by the activist shareholder Vestcorp (TFG Capital). A takeover of the target firm by the activist shareholder himself does not occur.

Table II.8. Percentage Size of Activist Stakes; The table shows the final sample of 253 observations. The percentage size is equivalent to the percentage of voting rights. There are no percentages available for seven observations. These percentages have been proxied with the help of information from corresponding news articles

Size of stake (in %)	Initial stake (Number of observations)	Maximum stake (Number of observations)
0 - 2.99	15	14
3.0 - 4.99	112	81
5.0 - 7.49	78	87
7.5 - 9.99	14	10
10.0 - 14.99	17	32
15.0 - 19.99	7	13
20.0 - 24.99	4	5
25.0 - 29.99	6	11
Minimum	0.10%	0.10%
Maximum	29.84%	30.33%
Arithmetic mean	5.94%	7.24%
Median	4.90%	5.09%

In at least 74 cases, the initial stake was later increased. In these cases the average size of the stake rises from 5.8 to 10.2 percent. Some increases might go unnoticed, however. Any investor increasing his stake from 3.0 to 4.9 percent or from 5.0 to 9.9 percent is not required to disclose his increase in voting rights (compare Table II.6.). If there is also no media coverage on the increase it will remain unnoticed. This leads to the conclusion that the actual number of increases is higher. From the investor's perspective an increased stake represents a stronger incentive to monitor the target firm's management. In addition, the investor's block of voting rights enables him to further influence voting outcomes at annual meetings with low attendance rates.

Table II.9. displays the amount of EUR invested throughout the events of the final sample. The average (median) target market capitalisation prior to the activist's investment is EUR 2,839 (360) million. The average (median) stake has a value of EUR 80.2 (17.2) million. In the case of the maximum stakes, the average (median) increases to EUR 96.0 (20.5) million. Twelve stakes have a market value of above EUR 500 million.

Potentially activist shareholders apparently have a preference for certain industries and industry conditions. One-third of target firms come from the machine construction sector. 17 percent of observations include firms from the financial services sector (banks, insurance companies, real estate).

Table II.9. *Value of the Activist Stakes in EUR Million; This overview is an approximation of values. The value is proxied by the target's market capitalisation at the end of the quarter preceding the investment. According to German law, the investor is not required to disclose his purchasing costs*

Percentile	Initial stake (in EUR million)	Maximum stake (in EUR million)
0.05	1.2	5.1
0.25	7.2	8.1
0.50	17.2	20.5
0.75	48.6	58.5
0.95	445.5	497.7
Minimum	0.28	0.34
Maximum	2628.4	2657.2
Arithmetic mean	80.2	96.0

Less than 20 percent of target firms have a negative return on equity. The median return on equity across the full sample is 10.4 percent. The average return on equity is similar for only the observations on Activity Level 3 and 4 (61 observations), as well as the hedge fund investor sub-sample. This looks quite different with respect to the stock price development of the target firms. More than half of the target firms have underperformed the market benchmark CDAX Performance Index in the twelve months prior to the potential activist's investment. The CDAX Performance Index is a broad, value-weighted market index. It consists of 600 German stock-market listed corporations. In comparison to the DAX and MDAX indexes, the CDAX also comprises firms with smaller market capitalisations. These findings would support the hypothesis that activist shareholders prefer healthy, profitable firms whose stock is underperforming the market. Debt-to-equity investments of potential activist shareholders were excluded from the analysis as described earlier. In the case of unprofitable firms in possible situations of financial distress investors may prefer to become shareholders by investing in bonds or bank loans to the target firm (see Brass (2010) for a discussion of this approach).

Table II.10. shows a classification of investor types. The investor with the single highest number of investments is former hedge fund manager Florian Homm (Absolute Capital Management Ltd.) who accounts for 32 potentially activist stakes. He is followed by Hermes, manager inter alia of the British Telecom's pension assets, accounting for 18 investments. Hermes has invested several EUR billions in German corporations including TUI AG, where the fund first invested in the 1990s. Hans-Christoph Hirt is Hermes's manager responsible for investments in Germany. Standard Capital Partners and well-known activist Guy Wyser Pratte (Wyser Pratte Management Co.) account for nine observations each.

Table II.10. General Classification of Potentially Activist Shareholders; Asset management firm includes mostly Anglo-American investment funds managing equity assets of at least USD 10 billion. Their funds are offered to the broad public and are therefore subject to stricter regulation. Pension funds are Hermes Fund Managers and the Teachers Insurance and Annuity Association (TIAA-CREF), a U.S. pension fund. Individual lists investments such as Rustam Aksenenko's stake in fashion company Escada. Private equity fund includes minority investments by private equity funds. All other observations are hedge fund investments. Hedge funds manage less than USD 10 billion in equity assets, do not offer fund investments to the broad public, run their funds from low-tax jurisdictions such as the Cayman Islands, use short-selling strategies and leveraging techniques, and are more short-term-oriented investors in many cases, especially compared to the pension funds

Investor type	Number of observations (full sample)	Number of observations (Activity Level 3 and 4)
Asset management firm	31	4
Hedge fund	182	45
Individual	8	2
Private equity fund	9	1
Pension fund	23	9

The analysis of the timing of the investment comes next. The German supervisory board is appointed for a fixed term of mostly five years. Do activist shareholders time their investment with respect to the next supervisory board election? This analysis requires information on the term of office of the next supervisory board. The information on the term of office of every single member of the supervisory board is not easily available. It is usually not presented on the company website and it is not contained in the annual reports. The commercial register (Handelsregister) shows only the names of the members of the supervisory board, but not until when they have been elected. This piece of information, however, is available from the annual meeting agendas of the target firms. The database of WAI Wirtschaftsanalysen und -informations GmbH contains meeting agendas of past annual meetings of publicly listed corporations. I use WAI database to determine every single supervisory board member's term of office.

Apparently, activist shareholders use a timing strategy with respect to the next supervisory board election. The average term of office of the supervisory board across the full sample is 4.8 years. This is close to the maximum possible term of office of five years. This is stipulated by section 102 sub-section 1 Aktiengesetz, the German stock corporation act. It will, henceforth, be abbreviated as AktG. New elections to the supervisory board are held just once in five annual meetings in the case of a five-year term.

Now, when comparing the date of the investment with the remaining term of office of the supervisory board, there seems to be a relationship. The frequency of activist stake-building increases when the next supervisory board election is moving closer. Table II.11. provides a corresponding overview. Theoretically, it is possible to influence the firm's management or the firm's strategy without the involvement of the supervisory board. Moreover, members of the supervisory board can step down before their term of office expires. This opens up the possibility for activist shareholders to promote the appointment of certain candidates. Nevertheless, the results shown in Table II.11. underline the relevance of the supervisory board election timing in German corporate governance.

Table II.11. Timing of Activist Investments; Number of observations is the number of investments during the respective time frame prior to the next ordinary supervisory board election. Number of observations (robust) excludes the following observations: (i) the term of office of the supervisory board is less than five years (23 observations), (ii) the supervisory board is staggered and more than just one election is necessary to replace all members (33 observations), (iii) the observation is based on a newspaper article and as a result the actual timing of the investment cannot be clearly determined (21 observations)

Time remaining until next supervisory board election:	Number of observations (full sample)	Number of observations (robust)
5 years	35	31
4 years	45	31
3 years	38	28
2 years	61	44
1 years	74	55

The Board of Directors of a U.S. corporation is elected at each annual meeting. The term of office of a German supervisory board is up to five years. This constitutes a major difference between German corporate governance and U.S. corporate governance. U.S. corporate governance in this context appears to be more short-term oriented. Many U.S. corporations have adopted a system of "staggered boards". Staggered board means that only half or only one third of board directors come up for election at each annual meeting (for a discussion of staggered boards see Bebchuk, Coates IV and Subramanian (2005)). This ultimately allows for a longer-term orientation in the work of the board of directors. International investors, especially Anglo-American investors, may not always be acquainted with the peculiarities of the German dualistic corporate governance system since it is quite different from the U.S. system in many details.

Target firm characteristics, the identity of activists and their investment approach have been discussed in this chapter. The next paragraph is an analysis of the activism itself and its effects.

II.5. Shareholder Activism and its Effects

Activity Level 1 (regulatory filing only) and Activity Level 2 investments (news mentioning) account for 192 out of 253 observations. The target firm and the capital market receive information about the arrival of a new shareholder, but not on his intentions. In these roughly 75 percent of cases there is no visible confrontation. A change on the supervisory board of the target firm does not occur. Conversations between the potential shareholder activist and the firm's management may still be taking place. And maybe the new shareholder is also starting talks with fellow shareholders. The point, however, is that there is no publicly available information on these instances.

Figure II.4. depicts the average daily trading volume in the shares of the target firms on the German stock market before and shortly after the announcement of a potentially activist stake. The average trading volume increases well before the announcement, especially within the [-150; -100]-event window. At least some activists build their stock positions by acquiring small blocks of stocks well in advance of the announcement.

As described above, the law stipulates a maximum cumulative number of 9 trading days between the acquisition of at least 3 percent of voting rights and the disclosure of the voting block to the capital market (section 21 sub-section 1 / section 26 sub-section 1 WpHG). The empirical analysis of disclosure filings shows that in some cases investors do not comply with this time frame. In case of stakes below the 3 percent threshold the investor can disclose his stake at any time. These remarks are important when interpreting the chart. The graph is an approximation for average daily trading volume of a target firm's stock.

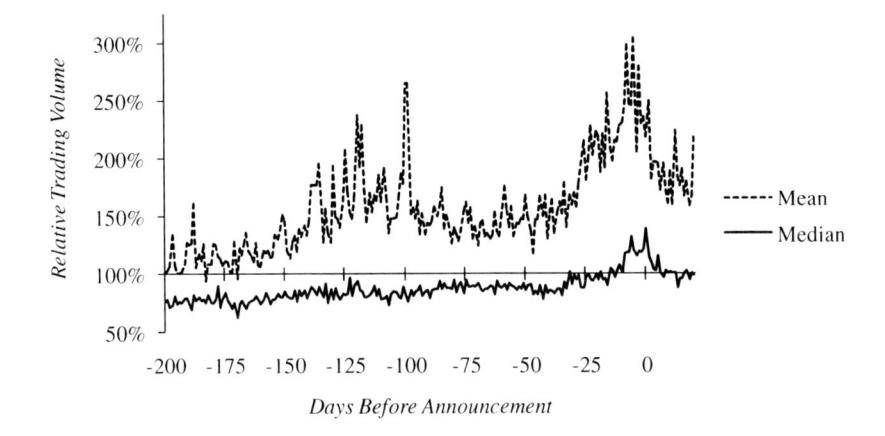

Figure II.4. Daily Trading Volume of Target Firms; "0" is the event day (day of disclosure of regulatory filing or disclosure of stake via news companies or the investor's website). The number of observations where the corresponding trading volume for the full 302 trading days surrounding the event was available is 224. Calculation: (i) for each stock the daily trading volume on all German stock exchanges in the [-282; -201]-window is determined; (ii) for each stock the daily trading volume in the [-200; +20]-window is determined; (iii) for each stock and each day a percentage is calculated using the formula ((ii) / (i))*100; (iv) a simple arithmetic mean for each day is calculated using the cross-section of 224 daily percentages from (iii). A result of 100% indicates no change in trading volume compared to the [-282; -201]-event window. Median shows the median value of the 224 daily percentages instead of the arithmetic mean. Source for stock price data: Thomson Datastream

The maximum daily trading volume occurs five days prior to the event date with a value of 304 percent while the maximum median value rests occurs on the event day itself. The maximum median value is 139 percent. This means that on the event day there is a *median* increase in trading activity of as much as 40 percent. The figure shows that an increased trading volume is a strong sign that an investor is accumulating a larger stock position.

The next table is looking at a corporate governance variable – a proxy variable for the level of shareholder monitoring, so to speak. Table II.12. shows the average annual meeting attendance rate before and after a potential activist shareholder acquires a stake. The average attendance rate prior to the investment is 50.27 percent. This is a value quite close to the average attendance rate for the DAX-30-listed corporations. For these corporations the average attendance rate is 53.04 percent (DSW e.V., 2008) between 1998 and 2008. Low attendance rates can be a threat to corporate governance and lead to random election results (Dauner-Lieb, 2007). There is no quorum requirement at annual meetings of German corporations (section 133 sub-section 1 AktG). A quorum requirement stipulates that a minimum percentage of shares, that is shares bearing voting rights, must be present at the annual meeting. Otherwise, no shareholder resolutions can be passed in that meeting. For U.S. corporations a quorum requirement generally exists. Not surprisingly, the lowest attendance rate in the constructed sample is 3.28 percent (Singulus Technologies AG, 2006 annual meeting). The investor relations department of CeWe Color Holding was able to increase the attendance rate at the firm's annual meeting from 38 percent in 2006 to approximately 87 percent in 2007. At that time activist hedge funds K Capital Partners and M2 Capital (the name was later changed to MarCap Investors) along with Guy Wyser Pratte opposed the firm's management.

The results shown in Table II.12. suggest that activist shareholders actively participate in annual meetings and cast their votes. The attendance rate significantly increases by an average 4.17 percentage points (p-value 0.03). The average initial activist stake of the activists is 5.94 percent. The rate of increase appears to be slightly below this value. However, when only including those activist stakes in the calculation where the respective attendance rates are available and meaningful the average initial stake decreases to 5.50 percent (159 observations). The two values, attendance rate increase and average initial stake, converge.

The average attendance rate for observations on Activity Level 1 and 2 is 51.37 percent. For observations on Activity Level 3 and 4 the average attendance rate is 45.77 percent. The attendance rate is approximately 5.6 percent lower on Activity Level 3 and 4. When comparing the median of the cross-section of attendance rates, the difference increases to 7.5 percentage points.

A shareholder who does not attend the annual meeting will probably not engage in any other monitoring efforts. Consequently, the annual meeting attendance rate of a corporation can be regarded as a proxy variable for level of shareholder monitoring, or vice versa, as a lack of shareholder monitoring. Following this assumption, the lack of shareholder monitoring is more pronounced on Activity Levels 3 and 4. The activist shareholders show an increased monitoring effort on Activity Level 3 and 4, including measures such as supervisory board representation (Level 4). This is consistent with the assumption stated above.

Both, average and median increases in the annual meeting attendance rate, are significant for the full sample as well as the two sub-samples, with the exception of the average increase on the combined Activity Levels 1 and 2. The strongest increase occurs on Activity Levels 3 and 4. The median increase, which is known to be a statistical measure robust against outliers, still shows a value of 5.5 percent and it is significant at the 5%-level.

Table II.12. Change in Annual Meeting Attendance Rates; Each annual meeting is considered only once. Subsequent annual meetings are considered only if at least one new potential activist shareholder has disclosed a stake. Annual meetings after the takeover of a target firm are not included. Difference-in-means test (Wilcoxon Signed Rank Test) displays t-statistics (z-score), significance level and in brackets the corresponding p-value for the differences in mean or median. *, ** and *** mark statistical significance 10%-, 5%- and 1%-level. Source for attendance rates is the WAI Wirtschaftsanalysen und -informations GmbH (www.hv-info.de) database as well as annual reports and target firm websites

	Attendance rate before investment in %	Attendance rate after investment in %	Difference-in-means test / Wilcoxon Signed Rank Test
Full sample (159 observations)			
Mean	50.27	54.45	2.16** (0,032)
Median	52.24	54.55	-3.35*** (0,000)
Activity Level 1 and 2 combined (128 observations)			
Mean	51.37	54.77	1.53 (0,127)
Median	53.16	56.47	-2.66*** (0,004)
Activity Level 3 and 4 combined (31 observations)			
Mean	45.77	53.11	1.99* (0,051)
Median	45.17	50.73	-2.03** (0,021)

Apparently, activist shareholders are active shareholders that participate in annual meetings and cast their votes. Language barriers for international investors can be overcome through appointing German-speaking fund managers or shareholder representatives. Examples for German speaking shareholder representatives are: Hans-Christoph Hirt (Hermes Fund Managers), Jens Tischendorf (Cevian Capital), Friedrich Merz (The Children's Investment Fund), Sebastian Freitag (M2 Capital and K Capital).

Table II.13. shows known cases where two or more activists have acquired a stake in the same target firm, a strategy also known as forming a "wolf pack". Through this simultaneous approach a single investor is able to share the risk associated with taking large stock positions in one enterprise with other investors. Furthermore, it saves an individual investor from reaching the 30 percent threshold of voting rights that triggers a mandatory takeover offer (section 35 WpÜG).

Table II.13. Simultaneous Investments by Multiple Activists; Voting block is the maximum percentage of voting rights cumulatively held by the activists. EURm is the estimated cumulative purchase price paid in the acquisition of the voting block assuming that the shares were bought at the market price. Both Voting block and EURm are estimations. Some funds are short-term investors that rapidly buy and sell their blocks or parts of their blocks within a few days or weeks. Not every trade triggers a mandatory disclosure. In the case of Demag Cranes several other activist funds acquired significant voting blocks after the announcement of a takeover by Terex Corporation

Target firm	Year	Potentially activist investors	Voting block	EURm
SGL Carbon	2004	Hermes, Jana Partners, K Capital, Marshall Wace	22.5%	86.2
Deutsche Boerse	2005-2006	Atticus, Harris Associates, Lone Pine, The Children's Investment Fund (TCI)	26.7%	1,515.6
Balda	2005-2007	Absolute Capital, Audley Capital, RIT Capital, Sapinda / Vatas, Wyser Pratte	42.6%	207.1
CeWe Color Holding	2005-2007	K Capital Partners, Lincoln Vale, M2 Capital (MarCap Investors), Seneca Capital, Standard Capital, Wyser Pratte	48.5%	129.4
Euromicron	2005-2006	Sapinda / Vatas, RIT Capital	29.7%	24.9
Curanum	2006-2007	Audley Capital, Global Opportunities Asset Management, Sapinda / Vatas, Wyser Pratte	40.1%	85.8
Praktiker	2007-2008	Artisan Partners, Bluecrest Capital, Eton Park Capital, Ivory Investment Management, Lansdowne Partners, Odey Asset Management	29.3%	351.5
Freenet	2007	Absolute Capital, Hermes, K Capital, Sapinda / Vatas	35.5%	697.1
Heidelberger Druckmaschinen	2007	Artisan Partners, Centaurus	8.1%	224.7
Demag Cranes	2010-2011	Centaurus, Cevian Capital, Elliott Associates	16.5%	91.9

Section 30 sub-section 2 WpÜG (acting-in-concert) is not to be violated. If a group of investors whose combined share in the company reaches 30 percent of voting rights unlawfully coordinates its activities, this also triggers a mandatory takeover offer. The acting-in-concert of multiple investors may be hard to prove in many cases.

Collaborating with other investors is apparently a frequently applied by some of the activists. The Swedish activist Cevian Capital serves as an example. The firm finances its investments purely with equity and can therefore only finance the acquisition of smaller stakes. However, it actively engages in talks with other shareholders to create a common understanding. This technique enables the investor to leverage its investment. This was also mentioned by Cevian Capital in the German financial newspaper Boersen-Zeitung dated 01 November 2011.

94 wolf packs occurred during the observation period. Two or more investors in the same company presumably have a similar schedule most of the time, but this is not necessarily the case all of the time. An example for diverging interests was the planned takeover of Hochtief AG by Spanish construction company ACS SA in 2010. While the U.S. hedge fund Southeastern Asset Management publicly urged Hochtief to accept the offer, pension fund Hermes did not support the takeover plans.

Further empirical evidence on the acting-in-concert of activist investors is not available from public sources.

The next chart, Table II.14., is more of a theoretical analysis. It shows how small voting blocks can have a significant impact in annual meetings. Individual minority shareholders can force or block certain shareholder resolutions if an annual meeting has a low attendance rate (see Dauner-Lieb (2007) for a general discussion of this problem in Germany). Theoretically, in the case of 30 annual meetings the target firm could have been forced to liquidate its business with a voting block of just 25 percent. Section 262 sub-section 1 AktG stipulates that three quarters of the votes cast at the annual meeting must approve the corporation's liquidation. Otherwise the resolutions will not be passed. At an annual meeting with an attendance rate of less than 33.3 percent one quarter of all outstanding voting shares account for three quarters of votes cast.

There is no visible confrontation between activist and the target firm's management in three quarters of all events. If a confrontation takes place, however, it can be quite intense. Public criticism towards the company is expressed in 41 cases. This involves 24 different target companies. The criticism includes demands for a new appointment of the supervisory board or the management board or a higher dividend payment as well as comments on the firm's mergers and acquisitions strategy.

There is a change on the target firm's supervisory board in 20 cases. These cases add to the 41 cases mentioned above and constitute all observations on Activity Level 3 and 4. In these 20 cases the change on the supervisory board is directly related to the activist shareholder. Some of these observations involve a public dispute, too. There are furthermore changes on the management board of four target firms attributable to the new shareholder among the 20 cases of supervisory board change. There are two events, where the activist failed to dismiss a member of the management board. In three cases, activists were successful in influencing the target firm's mergers and acquisitions strategy.

Table II.14. Possible Effects of low Attendance Rates in the Sample; Example: holding a block of voting rights of 10% a new supervisory board appointment could have been forced at five annual meetings; with a block of 15% this could have been achieved at 22 annual meetings. This example his hypothetical in nature since the supervisory board is not reelected at every annual meeting. The analysis takes into account attendance rates of 164 different annual meetings prior to the investment of a potential activist. Appoint new supervisory board refers to the members appointed by the shareholders in the case of codetermination. Data source for attendance rates is the WAI Wirtschaftsanalysen und -informations GmbH database

Block of voting rights	Block an amendment to the articles of incorporation (section 179 sub-section 2 AktG: 25% plus 1 share needed)	Appoint new supervisory board (section 101 / section 133 sub-section 1 AktG: 50% of shares needed)	Liquidation of target firm (section 262 sub-section 1 AktG: 75% of shares needed)
5%	5	1	1
10%	47	5	2
15%	114	22	5
20%	157	47	16
25%	164	76	30

Five extraordinary general meetings were held given the demands of activist shareholders. Three times a shareholder threatened to call an extraordinary general meeting. It seems that extraordinary meetings are actually a rare event since the full sample is comprised of 140 target firms and many activists hold their stakes for several years.

Interestingly, activist shareholders do not employ justice courts in relation with their investments, or their demands, respectively. One observed threat of legal proceedings occured, one case of a damage suit and four filed appeals against shareholder resolutions (Anfechtungsklagen). This does not include lawsuits related to a higher compensation of minority shareholders in the case of a squeeze-out (section 327 AktG) after the takeover of the target firm. Three of the four appeals filed are based on circumstances that commonly trigger appeals. As a comparison, the five most active plaintiffs in Germany have filed more than 32 appeals each according to the study of Baums et al. (2007). There are no signs that potentially activist shareholders force target firms into court proceedings as part of their strategy.

Table II.15. is a summary of major corporate events that occur within 18 months after the announcement of a potentially activist stake. 24 out of 140 target firms were acquired by a third party investor. 19 of the acquirers were strategic investors. Five acquirers were financial investors. A share buy-back was initiated in 16 cases. The share buy-backs were partly initiated in 2008 and 2009 – a time when the circumstances for initiating a share buy-back were generally favourable given falling share prices.

It can prove to be difficult to directly link the activist stake to a corporate event, that is, to *prove* causality. It is particularly noticeable, however, that 40 out of the 140 corporations have ceased to exist as independent entities, due to a takeover, liquidation or bankruptcy. A prominent example is the bankruptcy of Babcock Borsig AG in 2002, shortly after Guy Wyser Pratte had announced an activist stake in that firm. Newspaper articles claim that Wyser Pratte incurred a substantial loss on his investment.

Table II.15. Major Corporate Events at Sample Target Firms; Percentage of target firms is calculated on the basis of all 140 firms in the sample. The percentage on Activity Level 3 and 4 is calculated on the basis of 35 target firms and only counts the corresponding events. The corporate event is only considered if the activist shareholder was still invested in the firm at the time of the event. Takeover is the acquisition of at least 50% of voting shares by a third party investor after the activist's investment. As described earlier, 31 observations where an activist disclosed a stake after a takeover offer was announced were eliminated from the sample. Bankruptcy includes one case of nationalisation (Hypo Real Estate AG). Change on supervisory board includes cases where one or more seats have been obtained by the activist or a proxy close to the activist; more than one subsequent change can happen on a single supervisory board. Data source: company press releases and Lexis Nexis

Corporate event	Number of observations	Percentage of target firms	Percentage of target firms (Activity Level 3 and 4)
Takeover	24	17.1%	14.3%
Bankruptcy (liquidation)	15 (1)	11.4%	11.4%
Share buy-back initiated	16	11.4%	22.9%
Change on supervisory board	20	14.3%	48.6%

In less than one third of observations on Activity Level 3 and 4 the dividend payout ratio increases after the activist investment. The payout ratio actually declines in 41 percent of cases according to Thomson Datastream dividend payout data. Interestingly, the percentage of decreasing payout ratios is just 19 percent on Activity Level 1 and 2. One possible explanation is that activist shareholders abstain from criticising the target firm's management because of the more generous dividend policy.

The average holding period of the potential activist shareholders is hard to be estimated from publicly available information. The holding period reaches from several days to a decade and more. In most cases the exit date cannot be determined accurately. An investor is required to file a disclosure if his stake rises above certain thresholds. He is required to file a disclosure in the same manner if his stake falls below these thresholds. However, if an investor's stake falls below the 3 percent notification threshold this does not necessarily mean that he reduces his stake to 0 percent at the same time. Some evidence indicates that asset management firms and pension funds have longer holding periods, while hedge funds are more short-term oriented. Drerup (2010) in his study of hedge fund activism in Germany estimates an average holding period of 460 calendar days. Brav et al. (2008) indicate an average holding period of 22 months (approximately 660 calendar days) in their study of hedge fund activism in the U.S.

II.6. Summary of Results

The phenomenon of shareholder activism in Germany has been reviewed. The observation period started in January 1999 and reached until May 2011. A comprehensive data sample of 253 observations of potentially activist investments in 140 different target firms was constructed. The data was collected from various databases including the Bundesanstalt für Finanzdienstleistungsaufsicht (BaFin) database of significant holdings, the collection of electronic filings by the Deutsche Gesellschaft für Ad-hoc-Publizität and the database of WAI Wirtschaftsanalysen und -informations GmbH. 182 investments were sponsored by hedge funds, 31 investments were made by asset management firms and 23 of the potentially activist stakes were acquired by Anglo-American pension funds.

The average term of office of the target firm's supervisory board is 4.8 years. As the next ordinary supervisory board election moves closer, the possibility of a potential activist investment to occur increases. A part of the investors considers the timing of the next supervisory board election. A supervisory board seat offers the possibility for closer monitoring of the management.

The average attendance rate at the annual meetings of the target corporations prior to the activist investment is 50.27 percent. The sub-sample of observations on Activity Level 3 and 4 has a lower average attendance rate of 45.77 percent. The investment of a potential activist is followed by a significant increase in the average attendance of as much as 7.5 percent. This supports the hypothesis that activist shareholders actively engage in shareholder monitoring activities. In view of the low attendance rates of many corporations in the sample, activist voting can have a substantial impact on the outcome of shareholder resolutions despite their relatively small stakes.

75 percent of events are not associated with an open conflict or argument between shareholder and target firm and there is no change in corporate governance, that is, a change on the supervisory board or the management board. In those cases where there is a confrontation, activists try to push through their agendas. This includes taking seats on the supervisory boards of the target firms. Activists can leverage their investments through communication and cooperation with other shareholders. The empirical evidence suggests that this is a common strategy.

Overall, activist shareholders could have a powerful influence. The empirical evidence gathered suggests that the level of actual influence is rather moderate. Only a few German corporations were forced to adopt policies dictated by activist shareholders.

III. Activist Shareholders, Abnormal Returns, and the German Aufsichtsrat

Abstract

This event study finds significant, positive abnormal returns associated with stake-building by activist shareholders similar to those found in U.S. studies. I study the fixed term nature of the German supervisory board appointment hypothesizing that the timing of the upcoming election has an impact on the credibility of the activist effort. More credible approaches should consequently be associated with higher abnormal returns. An average abnormal return that is up to 6.9 percent higher can be earned in the [0; +5] event window when considering the timing of the next supervisory board election. Capital markets apparently perceive an activist effort within one to two years prior to the election as being most credible. Quite contrary to intuition the highest post-announcement abnormal returns can then be achieved with a comparably low stake. Talking about the credibility of the activist effort it seems that high cash positions on targets' balance sheets have a negative impact on the post-announcement abnormal return.

III.1. Introduction

Studies of German corporate governance focus on share ownership concentration (Franks and Mayer, 2001), large blockholders (Becht and Boehmer, 2003) and founding-family ownership (Andres, 2008), emphasizing the importance and influence of German banks ((Franks and Mayer, 1998), (Köke, 2004), (Heiss and Köke, 2004)). The influence of banks, however, has decreased in the past two decades ((Vitols, 2005), (Dittmann et al., 2010)) while regulatory initiatives have further increased accountability and transparency in corporate Germany ((Nowak, 2004), (Goergen et al., 2008)). Hackethal, Schmidt and Tyrell (2005) argue that the breakdown of the traditional system may lead to a control vacuum as a result of a growing lack of bank monitoring. This raises the following question: are there actually any owners monitoring German corporations?

In the unwinding of the so-called Deutschland AG banks including Commerzbank, Deutsche Bank, and Dresdner Bank as well as insurance companies like Allianz and Munich Re have sold off of their equity stakes in German corporations in the open market. Not surprisingly, more recent work documents the increased activity of international investors such as hedge funds and private equity funds ((Achleitner et al., 2010), (Bessler et al., 2010), (Drerup, 2010), (Achleitner et al., 2011), (Mietzner et al., 2011), (Rauch and Umber, 2012), (Drees, Mietzner and Schiereck, 2011)).

As the banks' grip on corporate Germany weakened, the average attendance rate at annual meetings declined. A compilation of data issued by the German proxy adviser Deutsche Schutzvereinigung für Wertpapierbesitz e.V. shareholder association (2008) shows that between 1998 and 2005 the average attendance rate at annual meetings of the DAX 30-listed firms dropped from 60.1 to 45.9 percent. It has rebounded since then but shareholders of German corporations are on average not active as almost one half does not attend the annual meeting.

A survey held by the German capital markets research association Deutsches Aktieninstitut e.V. (2009) has revealed that more than 70 percent of institutional investors would consider buying non-voting preference shares. While the survey relates the disinterest in voting shares partly to the presence of majority blockholders, it could as well be interpreted as an inclination towards shareholder inactivity. In addition, the language barrier and physical distance may impede international shareholders from actively participating in corporate governance, mainly the annual meeting.

The environment for activist shareholders in Germany is much more attractive now than it was ten to fifteen years ago ((Schaefer, 2007), (Goergen et al.,

2008)), resulting in increased investor activity as mentioned above. The purpose of this study, however, is not to come up with the latest and most comprehensive analysis of activist minority shareholders and abnormal returns in Germany. It is rather to explore whether some important elements of the German corporate governance framework may have gone unnoticed in the empirical literature so far.

Prior event studies on shareholder activism in Germany do exist. None of them, however, investigate the relationship between shareholder activism and the timing of the supervisory board election. Most studies apply models used in U.S. studies even though U.S. corporations do not have a supervisory board. The German corporate governance framework is therefore at the centre of attention of this analysis. This study relates the activist efforts to the timing of the next supervisory board election. With the election moving closer, abnormal returns tied to the announcement of activist stakes should be higher. Correspondingly, the frequency of activist events increases.

In the next paragraph three cases illustrate the role of the supervisory board when it comes to shareholder activism. This is followed by a literature overview and an explanation of the process of sample construction and the methodology applied. The analysis continues with a discussion of the event study results and a linear regression model explaining abnormal returns is applied. The study concludes with a review of findings.

III.2. Three Cases of Shareholder Activism

In January 2002, Guy Wyser Pratte disclosed a 5.01 percent stake in industrial conglomerate Babcock Borsig AG and announced an activist approach. The share price increased more than 20 percent within two days. Wyser Pratte openly criticised the management for taking the wrong decisions in the course of business. He claimed representation on the firm's supervisory board but was not successful in obtaining a seat. A few months later, Babcock Borsig faced liquidity problems and had to file for bankruptcy. Wyser Pratte incurred a substantial loss on his investment and later sued the firm's management for damages.

Among the most prominent examples of shareholder activism in Germany is Christopher Hohn's effort to block Deutsche Boerse AG's takeover of the London Stock Exchange through The Children's Investment Fund with the help of Atticus Capital and Harris Associates in early 2005, one and a half years ahead of the next supervisory board election. With a combined equity stake in Deutsche Boerse of more than 20 percent and the support of large asset management firms invested in the company, Rolf Breuer, chairman of the supervisory board, and Werner Seifert, Deutsche Boerse's CEO, were forced to step down. Deutsche Boerse's takeover effort was blocked and a large cash position was distributed to shareholders through share buybacks and dividends.

On 25 May 2010, Cevian Capital disclosed a 10.07 percent stake in Demag Cranes AG. Sweden-based Cevian Capital is known for having a track record as a shareholder activist in Europe. Demag's stock rised almost 14 percent in the 20 days surrounding the disclosure. Centaurus capital, a hedge fund based in London, disclosed the acquisition of a 3.37 stake on 18 October 2010. On 22 October, Jens Tischendorf, a representative of Cevian Capital was nominated to become member of the supervisory board after another member steps down. The new blockholders prompted Demag to start merger talks. Demag Cranes was ultimately acquired by U.S.-based Terex Corporation in August 2011. Within one year the value of Demag stock more than doubled creating substantial value for shareholders.

The three cases illustrate the relationship between the success of activist shareholder campaigns and corporate governance. The key takeaway from these cases is: the supervisory board is the pivotal corporate governance authority in Germany.

III.3. Literature Background, Corporate Governance Framework and Research Hypotheses

The modern corporation is characterised by the separation of ownership and control. While stewardship theory (Donaldson and Davis, 1991) depicts the manager as a "steward" of the company serving in the firm's best interest, agency theory (Jensen and Meckling, 1976) predicts that managers, who are not sole owners of the firm, will engage in activities that do not maximize the value of the firm. Jensen and Meckling define the concept of agency costs. Agency costs can arise from such things as perquisites (Yermack, 2006), entrenched boards (Bebchuk and Cohen, 2005) or entrenching investments (Shleifer and Vishny, 1989).

The empirical evidence on shareholder activism in its many varieties is vast. Karpoff (2001) and Gillan and Starks (2007) provide surveys of empirical findings, mainly for the United States. The most recent studies of shareholder activism by hedge funds and other entrepreneurial shareholder activists find significant, positive abnormal stock returns associated with the disclosure of an activist stake ((Brav et al., 2008), (Klein and Zur, 2009), (Greenwood and Schor, 2009)). In Germany, the most active investors both in terms of frequency and depths of activism are hedge funds, along with Hermes Focus, manager of, inter alia, the British Telecom's pension fund (see Becht, Franks, Mayer and Rossi

(2010b) for a clinical study of Hermes' activities in the UK). Larger and more institutional asset management firms, individuals and private equity investors taking minority stakes do occasionally engage in minority shareholder activism in Germany.

The attendance rate at annual meetings of German corporations has been relatively low for the past decade as described above. The lowest attendance rate within the sample constructed for the purposes of this study is at 3.28 percent (Singulus Technologies in 2006). According to the German stock corporation act there is no quorum requirement. Articles of incorporation (Satzung) can fix a minimum requirement, but most German corporations abstain from it. Additionally, bylaws (Geschaeftsordnung) set out by the supervisory board have a negligible meaning for shareholders of German corporations. Almost any change to the corporate governance architecture of a corporation only becomes effective through an amendment of the articles of incorporation, for which shareholders have to vote upon at the annual meeting (section 179 AktG). Management and supervisory board do not have the power to make amendments without consulting the annual meeting. This legal framework favours shareholder activism.

Any shareholder holding 5 percent of voting rights or EUR 500,000 of the share capital has the right to put items on an annual meeting's agenda (section 122 sub-section 2 AktG). Any shareholder resolution receiving a positive vote is binding in nature (Cziraki, Renneboog and Szilagyi, 2010). This is not the case in the U.S. If a supervisory board election is to take place, any shareholder has the right to submit nominations for the supervisory board election (section 127 AktG, full proxy access). There is an ongoing debate in the U.S. whether or not shareholders should have the right to initiate changes in the company's corporate governance (Bebchuk, 2005). As things stand now, shareholders of U.S. corporations do not have this right. A draft law by the Securities and Exchange Commission to give shareholders proxy access was overthrown in court in July 2011. Full proxy access is another supportive feature for minority shareholder activism in Germany.

In general, resolutions will be passed with a simple majority, meaning more than half of the valid votes cast, in the annual meeting (section 133 sub-section 1 AktG). However, the articles of incorporation can determine other majority requirements for elections (section 133 subsection 2 AktG), such as a second ballot, in which the relative majority shall be decisive. In practice, electoral majority requirements do differ.

According to the German corporate governance code (Deutscher Corporate Governance Kodex or DCGK) the election of the members of the supervisory board shall be conducted on a single member basis. Therefore an individual election for every single member may be held. The members of the supervisory board, however, usually share the same term of office.

The German corporate governance code is a set of non-binding rules on a comply-or-explain basis adopted in 2002 to promote standards of good corporate governance. If a member of the supervisory board steps down from office, the supervisory board or a shareholder, among others, are entitled to file a motion with the local court. A new member will then be appointed by the local court until the next annual meeting is held (section 104 AktG). Members of the supervisory board can be *removed* from office through a resolution of the annual meeting that requires 75 percent of the votes, unless otherwise provided in the articles of incorporation (section 103 AktG).

The management board and the chief executive officer are appointed by the supervisory board (section 84 AktG). Section 105 AktG prohibits a member of the management board of a German stock corporation from being a member of the supervisory board of the very same corporation. Members of the supervisory board must be non-executive, independent, outside directors or at most "gray" directors, that is having business relationships with the company (Bebchuk, Coates IV and Subramanian, 2002a).

A well-known and widely discussed characteristic of German corporate governance is mandatory co-determination on the supervisory board of most, but not all, larger German corporations ((Gorton and Schmid, 2004), (Fauver and Fuerst, 2006)). Labour representatives of either the workforce or labour unions fill board seats: in companies with more than 500 employees, one third of board seats, and in companies with more than 2,000 employees, half of the board seats. The latter situation is also called full parity, full co-determination or quasi-parity co-determination. This applies to most but not all industries. Industries exempt from mandatory co-determination include opinion-forming media companies and educational establishments such as universities. In the event of a tie of votes between labour representatives and shareholder representatives the chairman of the board has the power to decide on the respective issue (section 29 law of co-determination). On fully co-determined supervisory boards the chairman will be nominated by the shareholders (section 27 law of co-determination) while the labour representatives nominate the *deputy* chairman. However, it lies in the employees' own responsibility to organize the workforce and elect members to the supervisory board. If this is not done, there will be no co-determination – as is the case for a number of firms in the sample.

The maximum term of office of the members of the supervisory board is five years (section 102 AktG). Reappointment is permissible as well as usual (Hopt, 1997). But since the term of office of the members of the board of directors of a U.S. corporation is one year for an unstaggered board, the next election is always "right ahead" – not so in Germany. Franks and Mayer (2001) in their study on ownership and control of German corporations are among those who consider this difference:

"Since members of the supervisory board are appointed for fixed periods, it can take a considerable period of time for block holders to gain control of the supervisory board through new appointments."

The average term of office of members of the supervisory board in the cross-section of 253 events is 4.83 years, close to the maximum term of five years. Since the supervisory board is the pivotal authority in German corporate governance, gaining a seat on the supervisory board substantially increases the likelihood of success of any activist effort and therefore its credibility. Post-announcement abnormal returns should as a result be higher when the supervisory board election moves closer as the likelihood of success increases.

Given the existence of agency costs and the active approach of monitoring by new activist shareholders Hypothesis 1 claims:

H1: There is a significant, positive abnormal stock-price effect associated with the announcement of an activist minority stake.

With respect to the fixed-term nature of the German supervisory board's appointment Hypothesis 2 claims:

H2: Capital markets will perceive an activist effort by a minority shareholder within a time frame that is closer to the new supervisory board election as being more credible. Post-announcement abnormal returns are therefore higher for the respective observations.

III.4. Data and Methodology

There is no central database that stores the names of activist shareholders or activist events in Germany. I therefore form a list of potentially activist shareholders by gathering information from various sources including journal articles (for example Becht, Franks and Grant (2010a)) and by searching Bloomberg News and the Lexis Nexis database for articles on shareholder activism. For minority stakes below 10 percent an investor need not to disclose any of his intentions. For stakes between 5 and 10 percent there is no equivalent to the U.S. SEC 13D filing from which conclusions about the investor's approach could be drawn.

Using the names from this list I search five possible sources for mandatory filings of significant shareholdings by potential shareholder activists below the 30 percent threshold *and* that were not followed by a takeover of the very same investor. The five sources are BZ Pro, dgap.de, target websites, target annual reports and the financial markets authority Bundesanstalt für Finanzdienstleistungsaufsicht (BaFin). The BaFin database only lists current shareholdings. Hence, I recur to historic copies of the database. BZ Pro, hosted by the newspaper Boersen-Zeitung, and dgap.de are electronic archives of mandatory disclosures. The five sources are complementary. The number of stock market-listed companies in Germany is between 1,000 and 1,500. It is therefore possible to check almost every single company's announcements. Some companies have a minor free float thereby reducing the number of possible targets. I also look for announcements of activism only mentioned in the news to collect information about stakes below the 5 and 3 percent threshold, respectively. The threshold was lowered in 2007 from 5 percent to 3 percent. As Becht et al. (2010a) report this kind of process of data collection is quite straightforward and without convincing alternative. It yields a preliminary sample of 368 observations.

From the preliminary sample I exclude inter alia 31 cases of potential merger arbitrage (disclosure of stake after the announcement of a takeover bid by a third party), 30 events that occur within 282 trading days of the target's IPO (which would result in statistical issues), 20 debt-to-equity swaps (financial distress of target), ten cases where no event date could be found (in all ten cases the stakes were non-hostile, between 3 and 5 percent in size and acquired before and sold after January 2007), and eight observations of investments in non-voting preference shares (these eight events were collected from newspapers). The final sample consists of 253 potentially activist minority stakes in 140 target firms between January 1999 and May 2011, consistent with the number of events in the studies of shareholder activism in Germany by Bessler et al. (2010) and Drerup (2010).

I group each observation into one of four activity levels. Level 1 observations based on regulatory filings, Level 2 stands for regulatory filing and in addition mentioning in the news but without criticism, prerequisite for Level 3 is open criticism concerning the target's corporate governance and Level 4 means that there was actually a change on the supervisory board that can be attributed to the activist. Table III.16. describes the sample.

The grouping into four levels allows for a more differentiated analysis and it is still possible to consider hostile (Level 3 and 4) and non-hostile (Level 1 and 2) events separately. Even though I do not fully rely on media coverage, there may still be a bias towards larger companies in this sample (as can be seen from the mean and median EURm commitment figures on Activity Levels 2 and 3). Market capitalisation as an explanatory variable is for this reason not included in the analysis. Hostile in the sense of this study means increasingly active or confrontational. It is not meant in the sense of the event resulting in a hostile takeover of the target firm. Overall, the selected approach is the best possible match to Brav et al. (2008) and Becht et al. (2010a). With respect to the possibility of observing regulatory disclosures below the 5 percent threshold it may even be an improvement to Brav et al.'s approach.

Table III.16. Sample Description; Activity Level 1: regulatory filing, Activity Level 2 regulatory filing and newspaper mentioning but no criticism, Activity Level 3: criticism by activist shareholder directed towards target management which is mentioned in the news, Activity Level 4: change on supervisory board attributable to activist shareholder. For 7 observations below regulatory thresholds no %-stake size was available. The %-stake sizes for these observations were proxied by the average of possible sizes. EURm commitment is the value of the activist stake in EUR millions proxied by %-size of initial stake times the market capitalisation at the end of the quarter preceding the investment. N is the number of observations

Activity Level	Mean initial %-stake	Mean initial EURm commitment	Median initial %-stake	Median initial EURm commitment	Maximum initial %-stake	N
1	5.0	27.8	3.4	12.7	23.6	108
2	6.2	100.8	5.1	22.4	25.1	84
3	4.9	222.5	3.3	32.0	28.3	41
4	12.2	164.5	10.4	15.0	29.8	20
Full sample	5.9	94.4	4.9	17.2	29.8	253

Less than 25 percent of all events are hostile (61 observations). 20 actual changes on the supervisory board of 17 target firms were initiated by 15 different minority activist shareholders. The changes on the supervisory board may come a few weeks after the investment (Euromicron / Sapinda), a few months after the investment (Demag Cranes / Cevian Capital) or in some cases several years after the investment (Infineon Technologies / Hermes). The event date is always the date of the disclosure of the stake, even though the change on the supervisory board occurs at a later point in time.

The event study approach applied to measure abnormal returns is the same as in Achleitner et al. (2011) using the market model to calculate expected returns with the broad, value-weighted C-DAX performance index of approximately 600 German firms as a proxy for the market portfolio.

The event date is defined as the date of disclosure of the regulatory filing or, in case of a newspaper article, the date of publishing. Whenever I find two different dates I pick the earliest one. It took some time to assign the proper event dates as corrections of earlier regulatory filings happen to occur quite often.

An abnormal return of 5 percent for a firm that has a market capitalisation of EUR 500 million is equal to an immediate increase in shareholder value of approximately EUR 25 million. This is the link between agency costs, cumulative abnormal return and shareholder value.

III.5. Results

III.5.1. Announcement Effects

Table III.17. shows the average announcement effect on the share price of shareholder activist targets for the four activity levels across different event windows. The cumulative abnormal return on Activity Level 1 is below 1 percent across all event windows. Consistent with the Efficient Market Hypothesis (Fama, 1970) there is no difference between abnormal returns on Activity Level 1 and Activity Level 2. Once the capital market has knowledge of the potential activist investor's disclosure of a regulatory filing (Activity Level 1), a news article reporting on the very same disclosure (Activity Level 2) does not convey any new information to the capital market.

When shareholder activists take a hostile approach (Activity Level 3 and 4) cumulative average abnormal returns are larger. The mean cumulative abnormal return when combining Activity Level 3 and 4 in the event study is 4.38 percent in the [0; +5]-event window and it reaches 7.30 percent in the [-20; +20]-event window.

When comparing Activity Level 3 and 4 abnormal returns are higher on Level 4 reaching 11.28 percent in the [0; +10]-event window. This suggests that the anticipation of changes on the supervisory board (Activity Level 4) leads to higher abnormal returns. Obtaining a board seat increases the probability of success of the activist effort. The difference in abnormal returns between Activity Levels 3 and 4, however, could in part also be explained through size effects, as firms on Level 4 are on average smaller (see Table III.16.).

The size of the activist's initial stake seems to have an impact on abnormal returns, as the average initial stake on Level 4 is more than twice as high as on Level 3. This is consistent with the theory.

Table III.17. Announcement Effects on Different Activity Levels; Activity Level is described in Table II.7. Cumulative Abnormal Return is the sum of daily abnormal returns across the respective Event Window. Expected returns were calculated with the market model over the estimation period [t-282; t-30] with the C-DAX as market portfolio and the event date t=0. Boehmer Test as proposed in Boehmer, Musumeci and Poulsen (1991) is a modification of the traditional T-Test, which is robust towards event-induced variance. Wilcoxon Signed Rank Test is a non-parametric test (Wilcoxon, 1945) for difference-in-medians with the z-score being the standardised Wilcoxon test statistic. Share price data is from Thomson Datastream. ***, **, and * indicate statistical significance at the 1%-, 5%-, and 10%-levels, respectively

Event window	Cumulative Abnormal Return Mean	Median	t-Test t-value	Boehmer Test z-score	Wilcoxon Signed Rank Test z-score
Panel A: Activity Level 1 (108 observations)					
[-20;+20]	-0.18%	-1.39%	-0.10	-0.05	-0.37
[0;+0]	0.62%	0.14%	1.89*	1.80*	-1.33
[0;+1]	0.66%	0.28%	1.69*	1.32	-1.42
[0;+5]	0.65%	0.57%	0.94	0.71	-1.20
[0;+10]	0.42%	-0.62%	0.44	0.22	-0.07
Panel B: Activity Level 2 (84 observations)					
[-20;+20]	0.81%	2.70%	0.34	0.20	-1.29
[0;+0]	0.56%	0.12%	1.54	1.86*	-1.50
[0;+1]	0.37%	0.17%	0.81	1.30	-1.09
[0;+5]	-0.55%	-0.11%	-0.61	-0.40	-0.21
[0;+10]	-0.86%	-0.70%	-0.84	-0.68	-0.68
Panel C: Activity Level 3 (41 observations)					
[-20;+20]	4.36%	3.39%	1.88*	1.83*	-1.73*
[0;+0]	1.85%	0.86%	2.47**	2.87***	-2.44**
[0;+1]	2.28%	0.91%	2.71***	2.94***	-2.43**
[0;+5]	2.15%	0.75%	1.89*	2.00**	-1.39
[0;+10]	2.65%	1.87%	2.03**	2.17**	-1.67*
Panel D: Activity Level 4 (20 observations)					
[-20;+20]	10.13%	8.24%	1.74*	1.96**	-2.50**
[0;+0]	1.51%	0.37%	1.83*	2.08**	-2.20**
[0;+1]	4.31%	2.92%	2.81**	2.88***	-3.14***
[0;+5]	8.94%	3.28%	1.63	1.68*	-2.84***
[0;+10]	11.28%	5.82%	1.98*	2.04**	-2.99***

The results are robust to excluding target companies whose shares have an estimation window trading average of below 50 thousand shares per day on German stock exchanges. In cases of low stock market liquidity abnormal returns can partly originate from stock illiquidity. In 92 cases potential shareholder activists acquire a stake in a firm where another potential activist is already invested. This can lead to full or partial overlap of estimation windows with event windows of earlier observations thereby causing a potential bias in expected returns. When excluding the respective observations from the sample cumulative abnormal returns are slightly higher than reported in Table III.17. across all activity levels.

In order to facilitate further interpretation of the abnormal returns on Activity Level 1 and 2 an event study is conducted on 119 minority investments of non-activist institutional asset management firms including BlackRock, Fidelity Investments, Fidelity Management and Research, Schroder Investment Management, The Capital Group and Threadneedle. Results are shown in Table III.18.

When large, non-activist asset management firms disclose the acquisition of a stake an abnormal return in the [0; +1]-event window of approximately 0.7 percent can be observed. The magnitude of these abnormal returns is very similar to that of potentially activist events that are non-hostile (Activity Levels 1 and 2). This finding supports theories beyond agency theory. Superior stock picking ability may be the reason for the abnormal returns observed on Activity Level 1 and 2.

Table III.18. Announcement Effects of Non-Activist Investments; The table shows the announcement effects on the share price of German target corporations when non-activist institutional asset management firms disclose a stake. The number of observations is 119. For explanations of methodology and test statistics see Table III.17. Distribution of event dates across the observation period, characteristics of target firms and %-size of acquired stakes resemble those of the potential activist sample. **, and * indicate statistical significance at the 5%-, and 10%-levels, respectively

Event window	Cumulative Abnormal Return Mean	Median	t-Test t-value	Boehmer Test z-score	Wilcoxon Signed Rank Test z-score
[-20;+20]	-2.37%	0.11%	-1.711*	-1.564	-1.649*
[0;+0]	0.34%	-0.08%	1.313	1.309	-0.345
[0;+1]	0.71%	0.12%	2.389**	2.351**	-1.655*
[0;+5]	0.64%	0.01%	1.148	1.173	-0.339
[0;+10]	0.46%	-0.23%	0.604	0.424	-0.231

Overall, the results confirm Hypothesis 1 with respect to the hostile stakes (Activity Level 3 and 4). The results are also in line with the findings by Drees et al. (2011) who document positive, significant abnormal returns of up to 12 percent for activist blocks in Germany.

In cases where potential shareholder activists remain passive (Activity Level 1 and 2) significant positive abnormal returns can be observed. However, the results do not suffice to confirm Hypothesis 1 in the sense that these abnormal returns generate from the potential reduction of agency costs at the target firm by the activist shareholder given the results presented in Table III.18. (non-activist fund sample).

III.5.2. NewBET Analysis

The information on the term of office of the supervisory board members is not available from the articles of incorporation alone. The same is the case for various other sources such as the annual report. Studying the agendas and voting outcomes of target company annual meetings solves this problem.

I define *NewBET* as the *New Supervisory Board Election Timing*. NewBET can take on the values of 5, 4, 3, 2, and 1. Each value represents a time frame. NewBET 5, for example, applies to event dates within a time frame of more than four years and up to (the statutory maximum of) five years until the next supervisory board election. Another way to read the NewBET measure is taking it as the number of annual meetings until the next supervisory board election. NewBET 4 in this case means it will take four annual meetings for the supervisory board to come up for election. Capital markets at the time of the announcement of the activist stake will know that it will take a certain period, that is a certain number of annual meetings, for the supervisory board to come up for election. The NewBET analysis is presented in Table III.19.

The frequency of potentially activist events almost gradually increases from 35 events five annual meetings ahead of the supervisory board election (NewBET 5) to 74 events right ahead (NewBET 1) of the election as can be seen in Panel E. A comparison with the sub-samples in Panel F (excluding overlaps in estimation windows) and Panel G (hostile events only) confirms these findings. As a robustness check I exclude all events where the term of office of the supervisory board is less than five years (23 observations), all events where the target's supervisory board is staggered (33 observations) and all observations below regulatory thresholds (21 observations). There may be a more news on shareholder activism when the annual meeting is moving closer. The results remain the same. As a second robustness check I investigate the NewBET values for the sample of 119 investments by non-activist institutional asset management firms (see Table III.18. above). The NewBET distribution of these 119 events appears to be random. The modal value, that is, the highest value, is NewBET 3. These findings suggest that some activist shareholders apparently apply a timing strategy when engaging in activism, while non-activist shareholders do not time their investments.

Noteworthy at this point to mention, that the supervisory boards do not seem to be staggered for purposes of takeover defence. It rather looks like new members who fill vacancies are sometimes appointed for a maximum possible term of five years and not just for the remainder of the fixed term.

Since the frequency of events increases as the new supervisory board election moves closer there also happen to occur more overlaps in event and estimation windows towards the election. This creates an upward bias in expected returns. Overlapping events are therefore excluded from the analysis to achieve more robust results. The corresponding abnormal returns can be observed in Panel F.

The cumulative abnormal returns show different values for the five New-BET categories. Across all three panels cumulative abnormal returns are the highest at NewBET 2, that is, for events one to two years prior to the new supervisory board election. The abnormal returns reaches 5.29% in the [0; +5]-event window when observations with overlaps in the estimation windows are excluded.

Table III.19. Announcement Effects and NewBET; Example: NewBET 3 means it will take three annual meetings for the new supervisory board to be elected. In the case of 33 observations there was a staggered board; in these cases NewBET 3 means that it will take three annual meetings for more than half of the supervisory board seats filled by shareholders to come up for election. Test statistics for Panel G are partly omitted given the low number of observations. For further explanations see Table III.17. ***, **, and * indicate statistical significance at the 1%-, 5%-, and 10%-levels, respectively

	Cumulative Abnormal Return		t-Test	Boehmer Test	Wilcoxon Signed Rank Test
Event window	Mean	Median	t-value	z-score	z-score
Panel E: Full sample (253 observations)					
NewBET 5 (35 observations)					
[0;+0]	0.67%	0.13%	1.07	1.24	-0.59
[0;+5]	0.32%	0.74%	0.24	0.25	-0.97
NewBET 4 (45 observations)					
[0;+0]	0.93%	0.37%	2.65**	2.85***	-2.20**
[0;+5]	1.06%	0.76%	1.19	1.09	-1.01
NewBET 3 (38 observations)					
[0;+0]	0.48%	0.16%	1.14	1.08	-0.98
[0;+5]	-1.74%	0.23%	-1.11	-1.12	-0.17
NewBET 2 (61 observations)					
[0;+0]	1.47%	0.76%	2.74***	3.06***	-2.96***
[0;+5]	3.71%	0.75%	1.86*	1.95*	-2.04**
NewBET 1 (74 observations)					
[0;+0]	0.63%	-0.04%	1.27	1.35	-0.69
[0;+5]	0.97%	0.12%	1.12	1.15	-0.55
Panel F: Full sample excluding overlaps in estimation windows (177 observations)					
NewBET 5 (24 observations)					
[0;+0]	0.26%	-0.69%	0.32	0.40	-0.77
[0;+5]	-0.74%	0.53%	-0.40	-0.41	-0.17
NewBET 4 (33 observations)					
[0;+0]	0.98%	0.61%	2.28**	2.59***	-2.08**
[0;+5]	0.61%	0.32%	0.71	0.75	-0.49
NewBET 3 (29 observations)					
[0;+0]	0.79%	0.74%	1.51	1.42	-1.46
[0;+5]	-1.63%	0.70%	-0.86	-0.86	-0.29
NewBET 2 (45 observations)					
[0;+0]	1.73%	0.78%	2.43**	2.59***	-2.56**
[0;+5]	5.29%	1.58%	2.01*	2.08**	-2.50**
NewBET 1 (46 observations)					
[0;+0]	0.87%	0.04%	1.22	1.08	-0.30
[0;+5]	1.32%	-0.55%	1.10	0.81	-0.03
Panel G: Hostile events, [0;+5]-event window (61 observations)					
NewBET 5 (9 obs.)	0.83%	-0.65%			
NewBET 4 (7 obs.)	1.82%	1.06%			
NewBET 3 (7 obs.)	1.01%	2.63%			
NewBET 2 (18 obs.)	10.76%	3.92%	1.72*	1.80*	-2.51**
NewBET 1 (20 obs.)	2.31%	0.70%	1.55	1.46	-1.12

Figure III.5. depicts the relationship between NewBET and CAR using a bar chart. The focus here is on the [0; +5]-event window. The five-day period is long enough to capture the full effect of the announcement. On the other hand it is short enough to exclude post-event effects that are not related to shareholder activism. Such effects might deteriorate the results of the event study. Figure III.5. shows the median CAR and not the mean in order to account for outliers.

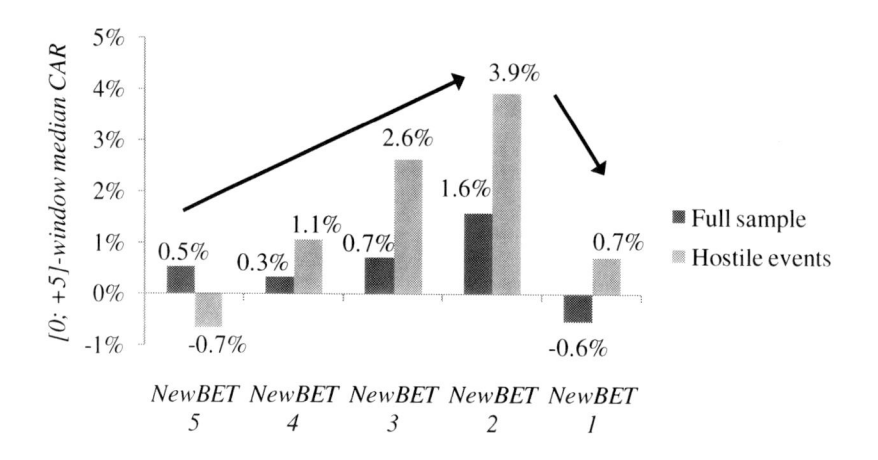

Figure III.5. Relationship Between Timing and CAR; The chart shows the full sample excluding overlaps in estimation windows (Panel F, 177 observations) and hostile events (Panel G, 61 observations)

Table III.20. exhibits a difference-in-means test for the mean cumulative abnormal returns observed in the [0; +5]-event window between NewBET 2 category and the four other categories. There is a significant difference between NewBET 2 and NewBET 3 as can be seen in Panel H of as much as 6.92 percent. Given the presence of outliers I winsorize each of the five distributions at the 5%-level. Panel I displays the results of the difference-in-means test between the winsorised samples. As expected, there is a significant difference between NewBET category 2 and categories 3, 4, and 5 taken separately. This confirms Hypothesis 2 to the extent that activist campaigns closer to the new supervisory board election tend to generate higher post-announcement abnormal returns. It does not support Hypothesis 2 in the sense that there is a strict inverse linear relationship between NewBET and abnormal returns. Looking at the median abnormal returns for the five NewBET categories for the hostile events (Figure III.5.), however, suggests that there is actually a partly linear relationship.

Table III.20. Difference-in-Means-Test; Mean CARs (cumulative abnormal returns) are displayed for the [0; +5]-event window. Observations with overlaps in estimation and event windows are excluded. Difference in means is the difference between the respective Mean CARs. T-stat and p-value are reported for the respective two-tailed difference-in-means test. N is the number of observations. Panel I shows results for the difference-in-means test when Mean CARs are winsorised at the 5%-level within each NewBET category. Winsorising (Dixon, 1960) changes the highest Mean CARs in a sample to the next smallest and the smallest Mean CARs to the next highest, thereby reducing the influence of spurious outliers without fully excluding them. Mean EURm market cap is the average target market capitalisation at the end of the quarter preceding the investment. **, and * indicate statistical significance at the 5%-, and 10%-levels, respectively

NewBET	Mean CAR	Difference in means	t-stat	p-value	N	Mean initial %-stake	Mean EURm market cap
Panel H: Full sample excluding overlaps in estimation windows (177 observations)							
5	-0.74%	6.04%	1.57	0.12	24	6.9%	1,441
4	0.61%	4.69%	1.48	0.14	33	6.9%	1,084
3	-1.63%	6.92%	1.91*	0.06	29	7.2%	2,897
2	5.29%	-	-	-	45	5.5%	2,348
1	1.32%	3.97%	1.39	0.17	46	5.7%	3,540
Panel I: Full sample excluding overlaps in estimation windows (177 observations) with distribution of mean CARs winsorized at the 5-% level							
5	-0.91%	4.42%	2.11**	0.04			
4	0.72%	2.79%	1.77*	0.08			
3	-1.00%	4.51%	2.46**	0.02	------------ unchanged ------------		
2	3.51%	-	-	-			
1	1.16%	2.35%	1.46	0.15			

The results are robust to applying a non-parametric test for unpaired samples, the Wilcoxon Rank Sum test (Wilcoxon, 1945). The median abnormal return in category NewBET 2 is 1.58% and the median abnormal return in the four other categories combined is 0.11%. The z-statistic for testing the null hypothesis that the two medians are equal is 1.97 with a one-tailed p-value of 0.024. The median abnormal return in the categories NewBET 3, 4 and 5 is 0.34% and the z-statistic testing the null hypothesis that the median equals the NewBET 2 median is 1.74 with a one-tailed p-value of 0.083.

Capital markets apparently perceive an activist effort within one to two years prior to the new supervisory board election as being most credible. By definition, *minority* shareholder activists who want to bring about change need to persuade fellow shareholders. In addition, ample communication may be necessary to convey the right information towards the target's supervisory board and management. Proxy proposals have to be prepared and submitted on time. All these efforts take time. Capital markets on average believe that less than one year is a very limited time frame as it seems. Contrary to intuition the highest post-announcement cumulative abnormal returns can be achieved with a comparably low minority stake. The mean initial percentage stake for NewBET 3, 4, and 5 is close to 7 percent while in NewBET category it is only 5.5 percent. Table III.19. and Table III.20. illustrate that both the highest mean and median abnormal returns were actually achieved at NewBET 2 with the lowest average percentage stake. This further supports Hypothesis 2.

III.5.3. Determinants of Abnormal Returns

Table III.21. presents an analysis of the drivers of the abnormal returns observed in the form of multivariate regression. The first model explains abnormal returns across the full sample. The second model, in addition to the first model, incorporates the annual meeting attendance rate of the last meeting prior to the event as an explanatory variable. The attendance rate of the annual meeting is probably among the best proxy variables for lack of shareholder monitoring. Shareholders that do not attend the annual meeting will in most cases not engage in any other monitoring activities. In fact, the average attendance rate increases by 4.2 percent (p-value 0.03) to 54.5 percent in the first annual meeting after the activist has disclosed his stake suggesting that activist shareholders actively participate in corporate governance. The third model explains the abnormal returns of the 61 hostile events (Activity Level 3 and 4).

The level of hostility of the activist approach, the percentage size of the activist's initial stake, and the timing with respect to the new supervisory board election (*NewBET 2*) have a significant effect on the post-announcement abnormal return.

Both, attendance rates (Model 2), and the level of co-determination (Model 3) do not have a significant influence on the post-announcement abnormal return. A reason for the non-significance could be the high correlation ($r=0.54$) between firm size as measured by market capitalisation and the variable full co-determination. There is generally a negative relationship between firm size and abnormal return ($r=-0.12$ for the full sample and $r=-0.27$ for Activity Level 3 and 4). Model 1 and 2 were estimated without the variable level of co-determination. Including this variable does not improve goodness of fit.

Even though the multivariate regression applying Model 3 indicates no significance of *Cofull* and *Cothird* it is interesting to see that despite the correlations described above both have a positive sign. Increasing the number of observations might lead to significant, positive values. This would then show that agency costs are actually higher at co-determined corporations.

A high cash position on the target's balance sheet seems to undermine the credibility of the activist effort. This can be seen from the respective negative coefficient in Model 1, Model 2, and Model 3. This result is in line with the findings of Bessler et al. (2010). It is robust to excluding financial services firms from the regression. Energy and utilities companies are not present in the sample.

F-statistics for all three models are highly significant. If I exclude all insignificant explanatory variables (*Wolfpack, Cothird* and *Cofull*) R-squared in Model 3 remains on the same level at 39.0 percent. This means the five remaining control variables have strong explanatory power in this Model. This includes the variable *NewBET 2*. *NewBET 2* is significant at the 5%-level and it also seems to have an economic effect given the comparably high coefficient in absolute terms of 0.041.

Table III.21. Results of Ordinary Least Squares Regression; The dependent variable is the [0; +1]-event window cumulative abnormal return. Activity Level 3 & 4, Wolfpack, NewBET 2, Cothird and Cofull are binary variables taking the value of 1 if the attribute is present in the given observation. Loginitialstake is the logarithm of the initial %-stake. Wolfpack is 1 if another activist is already invested at the time of the event. Prior12mperf is the target's share price performance relative to the C-DAX index in the twelve months prior to the event. RoE is the target's return on equity and CashtoAssets is target cash and cash-equivalents divided by total assets in the fiscal year prior to the event. AGM attendance rate is the attendance rate at the annual meeting prior to the event. Attendance rates are available for 209 observations. Cothird and Cofull stand for the level of co-determination on the target's supervisory board with Cothird meaning one third of the board seats are occupied by labour representatives and Cofull meaning half of the seats. Data source is Thomson One Banker for company financial data, Thomson Datastream for share price data and WAI Wirtschaftsanalysen und -informations GmbH for attendance rates and annual meeting agendas. Intercepts are suppressed because of the full span of dummy variables (Brav et al., 2008). T-statistics are shown in brackets and were computed using heteroskedasticity-robust standard errors (White, 1980). ***, **, and * indicate statistical significance at the 1%-, 5%-, and 10%-levels, respectively

	Model 1 All events (N=253)	Model 2 AGM attendance (N=209)	Model 3 Hostile events (N=61)
Activist approach			
Activity Level 3 & 4	0.024*** [3.244]	0.016*** [2.963]	
Loginitialstake	0.020*** [2.781]	0.023*** [2.718]	0.046*** [3.143]
Wolfpack	-0.011* [-1.905]	-0.006 [-1.198]	-0.013 [-0.965]
NewBET2	0.013** [2.422]	0.011** [2.286]	0.041** [2.421]
Target fundamentals			
Prior12mperf	-0.007 [-1.500]	-0.008** [-2.104]	-0.014* [-1.939]
RoE	0.004 [0.725]	0.003 [0.571]	0.034** [2.094]
CashtoAssets	-0.035** [-2.374]	-0.026** [-2.474]	-0.111*** [-3.765]
Corporate governance fundamentals			
AGM attendance rate	-	-0.014 [-1.455]	-
Cothird	-	-	0.011 [0.916]
Cofull	-	-	0.022 [1.665]
R-squared	15.8%	13.5%	43.3%
Adjusted R-squared	13.7%	10.5%	35.8%
F-statistic	3.87***	4.37***	7.59***

The size of the activist's initial stake has a significant, positive impact on the magnitude of abnormal returns, too. Larger activist stakes result in higher abnormal returns. This can once more be explained with arguments of credibility of the activist effort.

Including *pre*-announcement abnormal returns can bias results because for example larger stakes might cause larger pre-announcement stock-price run ups. The result is robust since the regression is based on *post*-announcement abnormal returns ([0; +1]-event window) and not on abnormal returns surrounding the event (for example [-20; +20]-event window).

All results are in line with the findings presented above supporting Hypothesis 1 as well as Hypothesis 2.

III.6. Conclusion

The analysis of regulatory filings and news articles yielded a sample of 253 investments by potential shareholder activists in 140 German publicly listed firms between January 1999 and May 2011. The sample includes 41 cases of open criticism towards the target's management but without any change on the supervisory board. It includes 20 cases of actual changes on the supervisory board attributable to activist effort. In sum, 61 of 253 observations were classified as hostile.

Abnormal returns are positive and significant for hostile as well as non-hostile events. Abnormal returns tend to be higher for hostile events and for events closer to the supervisory board election with the highest returns occuring within one to two years prior to the supervisory board election. All evidence suggests that post-announcement short-term abnormal returns are largely driven by the credibility of the activist effort to bring about change.

The empirical findings presented in this study provide an insight on how the German corporate governance framework affects the behaviour of capital markets participants and the stock price reactions tied to the announcement of shareholder activism. The results add to the picture painted by prior researchers on shareholder activism and shareholder monitoring.

IV. Weak Shareholder Rights? – A Case Study of Cevian Capital's Investment in Demag Cranes AG

Abstract

Activist shareholders Cevian Capital and Centaurus Capital played an important role in the takeover of Demag Cranes AG by U.S. rival Terex Corporation. This was possible through supervisory board representation and through publicly urging Demag to start merger talks. The Demag case shows that the German corporate governance system and minority shareholder rights are well-developed. It also serves as empirical evidence and that small shareholders actively participate in corporate governance. An overview of cases where seats on the supervisory board of German stock corporations have been obtained through minority shareholders is presented. This evidence further supports the strong shareholder rights perspective in Germany. The evidence presented in this study conflicts with the view of Shleifer and Vishny (1997): *"Germany (...) has virtually no participation by small investors in the market"* and the weak shareholder rights perspective in general.

IV.1. Introduction

The purpose of this study is to shed light on the topic of small investor participation and shareholder rights in Germany. Shleifer and Vishny (1997) in their survey of corporate governance around the world argue that *"Germany (...) has virtually no participation by small investors in the market"*. Is this true? The case study evidence presented in this analysis substantially challenges this view. Minority shareholder rights in Germany appear to be strong and small shareholders do participate in corporate governance. Differences between the German corporate governance system and its U.S. counterpart will be highlighted throughout this study as this is vital for understanding the discussion on shareholder rights in general.

Hellgardt and Hoger (2011) argue that there is a convergence between U.S. and German corporate law, but substantial differences between the two systems do remain. For example, the U.S. Court of Appeals in Washington, D.C. recently vacated a rule by the Securities and Exchange Commission (SEC), which would have required companies to include shareholders' director nominees in company proxy materials under certain circumstances. This can be learned from press statement 2011-179 by SEC chairman Mary L. Schapiro on proxy access litigation dated 06 September 2011. Shareholders of a German stock corporation (Aktiengesellschaft, AG) do have the right to make nominations for the supervisory board election (section 127 AktG) and the nominee will then come up for election at the annual meeting.

Another essential distinction between the two systems is in its board system. German corporations have a two-tier governance system with the board being divided into management board (Vorstand) and supervisory board (Aufsichtsrat) as a matter of law (Hopt, 1997). Many other key disparities but also similarities exist and they will be discussed later. It is well-known that U.S. law is based on common law while Germany has a civil law system. And it is an easy to underestimate challenge for researchers in the field of corporate governance and financial economics to completely consider these differences when it comes to details (Fleischer, 2008).

The case of Cevian Capital's investment in Demag Cranes in May 2010 serves as an illustrative example for the purposes of this study for three reasons. First of all, Cevian Capital acquired a minority stake in Demag Cranes of approximately 10 percent of voting rights, which is neither above the threshold for formal control nor sufficient voting power for gaining factual control. Secondly, Cevian Capital –as a minority shareholder– obtained supervisory board representation within five months of its investment. This suggests that the investor intended to participate in corporate affairs in the first place. A prerequisite for this intention is the activist's perception that this is possible. Thirdly, fellow minority shareholder Centaurus Capital showed active participation in corporate affairs by advocating a sale of the company. This makes the Demag case an incidence of activity by multiple small shareholders. Demag Cranes was eventually acquired by a competitor. This created value for all shareholders. The Demag case presents meaningful evidence for the strong shareholder rights perspective in Germany.

Cevian Capital's mandatory regulatory disclosure according to section 27a WpHG will be reviewed as part of this study. Section 27a WpHG can be seen as the equivalent to the U.S. SEC 13D filing, which requires an investor to disclose his intentions for stakes of 5 percent of the voting shares or above to the capital market. The mandatory disclosure according to section 27a WpHG was introduced in Germany in 2007 to promote transparency and to protect small shareholders in cases where investors acquire 10 percent of voting rights or more. Becht (1997), more than a decade ago, emphasized the need for European mandatory voting rights disclosure. Today, Germany has mandatory disclosure rules in place. The market has become more transparent, especially for small investors.

There was a series of influential papers by La Porta, Lopez-de-Silanes, Shleifer, and Vishny ((1997), (1997), (1998), (1999), (2000a), (2000b), (2002), (2006)) on the relationship between corporate ownership, corporate governance, investor protection, (minority) shareholder rights, and securities law around the world. These studies took a comparative approach pointing out differences between common-law countries and civil-law countries. These studies are among the most cited papers in the field of international corporate governance and financial economics research. According to the Google Scholar search engine the above mentioned publications have a combined 35,000 scientific and non-scientific citations as at November 2012.

Common to La Porta et al's research is that –as far as Germany is concerned– the studies were conducted in an era of strong influence by banks and insurance companies through equity ownership, proxy voting provisions and strong creditor relationships that favoured banks. In addition, La Porta et al. on a country-basis used small samples of only the largest firms (for example the 20 largest firms per country in the case of La Porta, Lopez-de-Silanes, and Shleifer (1999)) to draw conclusions about ownership patterns of all firms in this market. This research design led inter alia to the finding that German corporations frequently have controlling shareholders sharing their control only with the banks involved. La Porta et al. in their various studies concluded weak shareholder rights opposed to strong creditor rights, no protection of (minority) shareholders and no participation by small investors in German-civil-law countries including Germany itself.

Kim, Kitsabunnarat, and Nofsinger (2007) examine minority shareholder rights in Europe and apply a similar research design that leads to comparable findings with respect to the weak shareholder rights perspective. Their sample consists of 229 firms from across Europe. The German sub-sample consists of 23 of the largest German corporations. This limitation was chosen since corporate governance information regarding the corporation's directors was not available for smaller firms:

"Our source for firm's board information is Deminor (...) they only track the larger European firms. (...) We realize that prior cross-country studies (...) usually (...) include thousands of firm-level observations, but information on director independence on this scale is unavailable in machine readable form. (...) The countries' minority shareholder rights and legal enforcement measures used in our study come from La Porta et al. (1998)."

This quote emphasizes that (i) the findings by Kim et al. (2007) are based on samples that include only the largest German firms and may therefore not serve as a proxy for the German stock market as a whole and (ii) recent research of German corporate governance is still based on La Porta et al.'s findings of the 1990s.

Drerup (2010) analyses hedge fund activism in Germany and also follows this traditional view on the level of minority shareholder rights in Germany:

"I suggest that the comparatively high level of ownership concentration in Germany in combination with rather weak minority shareholder rights prevents shareholder activism from being effective."

Drerup differentiates between ownership concentration and minority shareholder rights. There are in fact certain stock market segments in Germany where ownership concentration is effectively higher compared for example to the U.S. market. The *weak* minority shareholder rights view, however, is outdated. The research at hand provides evidence in support of the *strong* minority shareholder rights view.

The analysis proceeds as follows. In the next paragraph emphasis will be laid upon the comparison of the supervisory board and the board of directors, which can be seen as its U.S. counterpart. This is followed by an introduction of the transaction partners Demag Cranes and Cevian Capital. In the main section Cevian Capital's investment in Demag Cranes is discussed including a short review of its follow-up investment in Bilfinger Berger SE. An overview of cases of supervisory board representation by minority shareholder activists in Germany is provided and discussed in the subsequent paragraph. This is followed by a summary of results and an outlook.

IV.2. German Supervisory Board and U.S. Board of Directors

Table IV.22. represent s a direct comparison of the German supervisory board and the U.S. Board of Directors. The details shown in the table will be explained throughout this section. Outside, or non-executive, directorship (Hopt, 1997), labour co-determination (Gorton and Schmid, 2004) and appointment for fixed periods (Franks and Mayer, 2001) of mostly five years are the attributes that best describe the supervisory board of a German stock corporation compared to the board of directors of a U.S. corporation. More differences exist. For an in-depth discussion of the German supervisory board from a comparative corporate governance perspective see Hopt (1997) and Hopt & Leyens (2004).

Inside directorship is prohibited since a member of the management board cannot be a member of the supervisory board of the same firm (section 105 subsection 1 AktG). U.S. corporations have a one-tier governance system consisting of the board of directors. Directors may be outside, independent directors or inside directors. In many cases the chief executive officer of the corporation acts as the chairman of the board of directors.

In Germany, the legal rules governing the supervisory board are mandatory and by-laws as set out by the supervisory board itself have a negligible meaning from a shareholder rights perspective in most cases. This is different for U.S. corporations where it may only be the by-laws governing such things as newly created directorships: compare for example article 7 of the certificate of incorporation of Terex Corporation filed with the Securities and Exchange Commission on 16 February 1994 and article 3.2. of the by-laws of Terex Corporation dated 05 March 2008.

Significant changes in most corporate control provisions of German stock corporations merely become effective through an amendment of the articles of incorporation (Satzung; the two expressions articles of incorporation and certificate of incorporation are used synonymously in the literature). It is an irrevocable shareholder right of the shareholders to vote upon any amendment of the articles of incorporation at the annual meeting (section 179 AktG) requiring at least 75 percent of votes cast in order to be passed. Management and supervisory board only have the power to *propose* amendments to be voted upon at the annual meeting.

A German equivalent to the U.S. Delaware General Corporation Law does not exist. Any German stock corporation is subject to the mandatory rules of the Aktiengesetz (German stock corporation act). The Delaware General Corporation Law (DGCL) is applicable to firms incorporating in the U.S. state of Delaware. DGCL can be characterised by stronger management rights and weaker shareholder rights (Cary, 1974). Kirk (1984) discusses the early evolution of corporate law in the State of Delaware in more detail.

Table IV.22. Comparison of U.S. Board of Directors and German Supervisory Board

	Board of directors (U.S. corporation)	German supervisory board (Aktiengesellschaft)
Nomination and election of directors		
Executive directorship	Yes	No
Cost of including nominee in proxy materials in any case borne by corporation	No	Yes
Election of directors without majority of votes at annual meeting possible	Yes	No
Quorum requirement at annual meeting	Yes	No
Board structure		
Labour co-determination for corporations with more than 500/ 2000 employees	No	Yes
Appointment of members for fixed period	No	Yes (up to five years)
Newly created directorships by change of by-laws (Geschaeftsordnung)	Yes	No
Other similarities and disparities		
National stock corporation act is mandatory	No (Delaware General Corporation Law)	Yes
Rules in place to promote transparency and good corporate governance	Yes (Sarbanes-Oxley Act)	Yes (Deutscher Corporate Governance Kodex)
Increase share capital without prior approval of shareholders	Yes	No

A German stock corporation with more than 500 employees is subject to the laws of co-determination. Full co-determination on supervisory boards is applicable for firms employing 2,000 or more employees. In the event of a tie of votes between labour representatives and shareholder representatives the chairman of the board has the power to decide on the respective issue (section 29 law of co-determination). On fully co-determined supervisory boards the chairman will in almost any case be nominated by the shareholders (section 27 law of co-determination). Controversial issues to be decided upon by the supervisory board are likely be decided by the shareholder representatives vote. This is an important feature to understand when talking about co-determination and shareholder rights in Germany.

There is generally no quorum requirement in Germany. At the annual meeting resolutions are passed with a simple majority of votes cast provided the articles of incorporation do not foresee a quorum requirement (section 133 sub-section 1 AktG). The articles of incorporation of both Demag Cranes and Bilfinger Berger do not foresee a quorum requirement. While German corporate law is largely based on the expectation that shareholders physically appear at the annual meeting, U.S. corporate law is known for its proxy voting provisions (Hellgardt and Hoger, 2011). Proxy voting is the process of delegating votes to another person (proxy) who then votes on the respective issue in the absence of the holder of the votes (principal) (Hofstetter, 2008). The purpose of the proxy voting system in a country of large size such as the United States is that more votes will be cast at the annual meeting.

Resolutions at the annual meeting of German corporations are passed with a simple majority of votes, that is 50 percent of the votes cast (section 133 sub-section 1 AktG). With respect to elections, the articles of incorporation can provide different requirements (section 133 sub-section 2 AktG). As discussed above, shareholders generally have the right to make nominations for the supervisory board election (section 127 AktG) and the nominee will then come up for election at the annual meeting. The cost associated with including the nomination of the shareholder's candidate in the corporations' agenda for the annual meeting will in any case be borne by the corporation and not the shareholder (section 126 AktG). Shareholders of U.S. corporations will be faced with the cost of including a nominee in the corporation's proxy materials if their candidate is not elected. This was decided in the case of Rosenfeld versus Fairchild Engine & Airplane Corporation, 128 N.E.2 d 291 (N.Y. 1955) (so-called "Froessel-Rule").

In the case of Terex Corporation, incorporated in the State of Delaware, plurality voting according to section 216 sub-section 3 DGCL applies. This can also be learned from article 2.8 of the by-laws of Terex Corporation. Absent a contested election, directors can therefore be elected with the vote of a single share cast in favour of the nominee. In the U.S., directors are nominated by the incumbent board of directors and according to SEC rule 14a-8 (i) the corporation has the right to turn down shareholder proposals that relate to the nomination of directors (Hellgardt and Hoger, 2011). Considering that the SEC was not successful in changing the respective regulation as mentioned in the introduction, activist shareholders in the U.S. face a difficult environment for the nomination and election of the board of directors. In comparison, the electoral procedure of the German supervisory board offers better possibilities to activist shareholders.

The maximum term of office of the members of the German supervisory board is five years, or in legal terminology five accounting periods (section 102 sub-section 1 AktG). Generally, the purpose of such a comparably long period is to provide management with a certain degree of security and independence in office for a number of years (Hopt, 1997). In the U.S., the board of directors usually have a one-year term and elections will be held at each annual meeting. Prominent examples are the boards of directors of Exxon Mobil Corporation, General Electric Corporation, Microsoft Corporation as well as Terex Corporation.

Annual elections may result in constant efforts by activist shareholders or potential acquirers to obtain board seats. Bebchuk and Cohen (2005) point out that the majority of U.S. public companies have a staggered board as opposed to a unitary board with one class of directors standing for election at every annual meeting. A staggered board typically has three classes of directors with only one class standing for election at each meeting. This can be seen as an inclination towards creating a more stable environment for management and a less favourable environment for shareholders (Bebchuk, Coates IV and Subramanian, 2002b).

In their research on the ownership and control of German corporations, Franks and Mayer (2001) remark that it can take a considerable period of time for a new blockholder to gain control of the supervisory board through new appointments given a maximum possible term of five years.

In 2002, at the time when the Sarbanes-Oxley Act came into effect in the U.S., the Deutscher Corporate Governance Kodex (DCGK) was established. The DCGK is a Corporate Governance Code that was adopted in Germany. It is a set of recommendations and suggestions on a comply-or-explain basis to further promote transparency and standards for good and responsible corporate governance. Certain parts of the DCGK can have a material impact on the outcome of a takeover effort.

The German stock corporation act as well as the German corporate governance code also provide rules on such issues as the maximum number of supervisory board mandates for a single person or prohibition of cross-directorships (compare section 100 sub-section 2 no. 3 AktG) and recommendations on the formation of supervisory board committees (article 5.3 DCGK).

Article 3.7 of the DCGK provides that

"After the announcement of a takeover offer, the Management Board may not take any actions outside the ordinary course of business that could prevent the success of the offer unless the Management Board has been authorized by the General Meeting or the Supervisory Board has given its approval."

The management and the supervisory board of the target are bound to the best interests of the shareholders and the enterprise. In the U.S., especially in the case of hostile takeovers, the board of directors of the target may without prior approval of shareholders issue new shares and dilute the bidder's block of shares (Hellgardt and Hoger, 2011). Empirical evidence exists that target shareholders often try to remove these takeover-blocking "poison pills" through shareholder proposals (Bizjak and Marquette, 1998). Issuing new shares without the prior approval of shareholders is not possible for German stock corporations. If a prior approval (Vorratsbeschluss) was given, the management has the right to issue new shares in line with the respective shareholder resolution. Demag Cranes had such an approval in place. The management board was authorised to increase the firm's capital by up to 50 percent.

The results from the above comparison of the U.S. board of directors and the German supervisory board barely support the weak shareholder rights view for Germany. Rather it seems that the U.S. board of directors is a body that operates more independently and that has the ability to largely control more many decisions that affect shareholder rights. Bebchuk (2007) describes the reality of corporate elections in the U.S. as a "myth of the shareholder franchise".

When evaluating the Demag case and the minority shareholder rights exercised by Cevian Capital several things should be kept in mind: (i) the supervisory board is appointing and overseeing the management; (ii) it is required to act in the shareholders' best interest; (iii) the supervisory board is involved in receiving and evaluating takeover offers for the firm. From a shareholder's perspective it is therefore the key authority when it comes to influencing the governance and decision-making of German corporations.

IV.3. Transaction Partners

Demag Cranes is a provider of industrial cranes and harbour cranes with a turnover of EUR 931 million for the accounting year ending 30 September 2010. The firm is head-quartered in Dusseldorf, Germany. Demag Cranes was sold by its owners private equity firm Kohlberg, Kravis and Roberts and Siemens AG in an initial public offering in June 2006 ultimately reaching a free float of 100 percent. Table IV.23. introduces the company in more detail.

Demag Cranes had been profitable before, during and after the financial crisis of 2008 that caused an unprecedented decline of industrial goods orders and consequently turnover in various sectors. The firm's main competitors were Konecranes Oyj from Finland with an annual revenue for the fiscal year 2010 of EUR 1,546 million and U.S. firm Terex Corporation with an annual revenue of USD 4,418 million (EUR 3,333 million).

Table IV.23. Company Description and Market Information; Company financial figures are shown in EUR millions, except earnings per share and share price (shown in EUR). EBIT stands for earnings before interest and tax. Share price shows the reporting date closing price. KKR stands for U.S. private equity investor Kohlberg, Kravis and Roberts. Data source is the Demag Cranes annual report 2009/10. The number of shares outstanding is 21,172,993. Additional source for ownership structure is Boersen-Zeitung

	2005/06	2006/07	2007/08	2008/09	2009/10
Revenue	986.9	1,080.4	1,225.8	1,047.6	931.3
EBIT	54.1	82.0	135.8	13.2	50.2
Number of employees	5,680	5,813	6,093	5,934	5,711
Ownership structure (before IPO: KKR / Siemens)	Free float 57%, KKR 32%, management 11%	KKR sells stake to institutional investors in December 2006; the management's stake drops below 3% in January 2007; free float ultimately reaches 100%			Cevian Capital discloses 10.07% stake
Market capitalisation	571.7	699.6	589.2	519.6	595.8
Earnings per share	1.04	1.53	3.79	0.04	1.31
Share price	27.00	33.04	27.83	24.54	28.14

Due to its strong business model and a free float of Demag's shares of close to 100 percent Demag could have become a takeover target at any point in time. Before May 2010, however, there was no serious bidder. This was also reflected in the weaker share price compared to the previous two fiscal years (see share price information shown in Table IV.23.).

At the end of May 2010, Cevian Capital disclosed a 10.07 percent stake in Demag Cranes making the investor the single largest shareholder. At that point in time there was no other shareholder holding a stake larger than 5 percent in the firm. Cevian Capital is a Sweden-based fund manager founded in 2002 by Christer Gardell and Lars Förberg. The investment funds managed by Cevian

Capital are set up as limited partnerships with the limited partners being insurance companies and other financial institutions. The managing partner Cevian Capital participates in capital gains.

The strategy of the fund is not to acquire whole firms like a conventional private equity fund. Instead it purchases significant minority stakes in companies (Sunesson, 2008). This investment strategy results in Cevian Capital's highly concentrated portfolio with only a few large investments. Correspondingly, the fund manager has a strong incentive to maximise the value of every single investment. Cevian Capital tries to achieve value maximisation by actively influencing the target firm's management and the general course of business. Bethel, Liebeskind and Opler (1998) investigate the causes and consequences of activist block share purchases labelling this as part of the market for partial corporate control. Cevian Capital is active in the market for partial corporate control.

84

Figure VI.6. Annotated Share Price Graph of Demag Cranes AG (ISIN DE000DCAG010) Between 01 January 2010 and 31 July 2011: The DAX-30 Performance Index (Reuters Symbol: .GDAXI), is used as a benchmark index. It is comprised of 30 of the largest and most actively traded stock corporations in Germany. Share prices and index values are rebased to the value of 100 on 01 January 2010. The illustration shows daily closing prices. Source for share price data and index values is Thomson Datastream

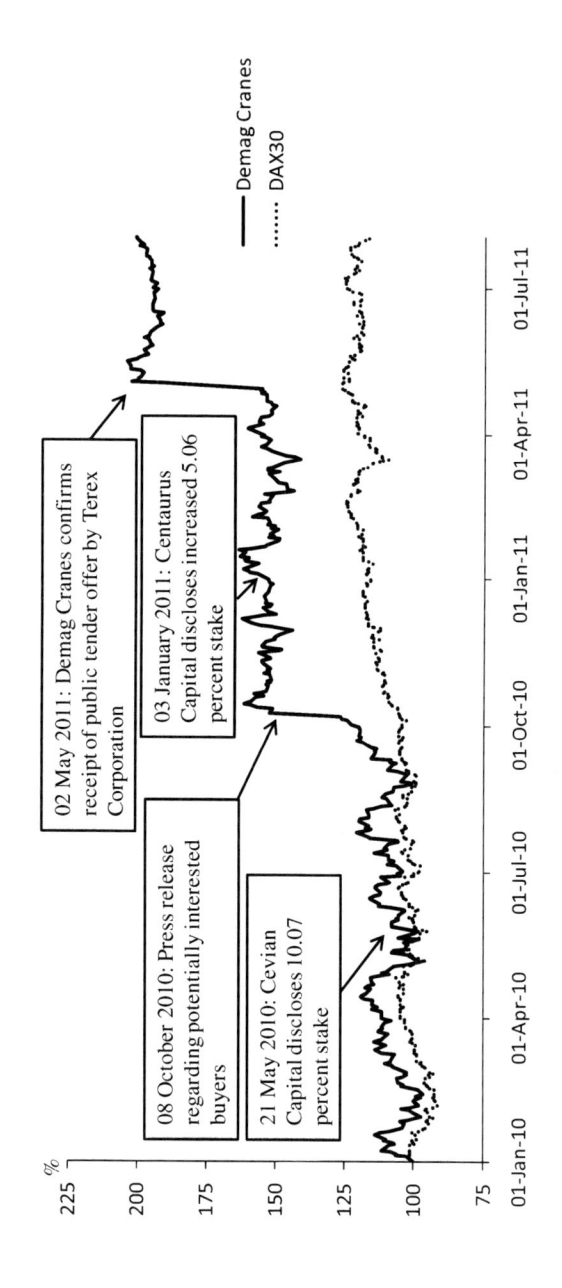

So far, Cevian Capital's investments in Germany to grab the attention of capital markets were the acquisition of a stake, first in insurance company Munich Re in 2007, and then in car manufacturer Daimler in 2008. Both stakes had been below the 3 percent threshold meaning no regulatory disclosure was filed. However, in both cases there were press articles reporting on the new shareholder Cevian Capital.

Figure VI.6. depicts the development of Demag's share price between 01 January 2010 and 31 July 2011. The share price increased during this period both in terms of absolute value and in terms of relative value. The acquisition of Demag Cranes by Terex Corporation ultimately created significant value for shareholders.

IV.4. Investment by Cevian Capital and Takeover by Terex Corporation

IV.4.1. Investment by Cevian Capital

On 21 May 2010, Demag Cranes issued a press statement on the acquisition of a stake in the firm by Cevian Capital. The disclosure stated a participation of slightly above 10 percent making the fund the largest shareholder. In a corresponding press statement Cevian Capital deemed Demag Cranes as a fundamentally undervalued company. On 25 May 2010, a stake of 10.07 percent was officially disclosed. A formal regulatory filing according to section 27a WpHG was issued through Demag Cranes stating that (i) the investment by Cevian Capital was made for trading and not strategic purposes, (ii) further shares may be acquired but no more than 25 percent of voting rights in total, (iii) supervisory board representation would be sought but no change on the management board was planned, (iv) no changes to the capital structure or the dividend policy would be proposed, and (v) the purchase of the shares had been financed with equity.

The details of the section-27a-WpHG statement reveal Cevian Capital's intention tied to the equity stake. The information provided indicates the active stance of the investor. It appeared that Cevian Capital wanted to participate in Demag's corporate governance despite being just a minority shareholder.

Cevian Capital's investment had been financed with equity. Some international investors, for example many hedge funds, employ debt to finance their investments. These investors often borrow against the shares purchased. Some hedge funds propose higher dividend payments upon their investment or at a later point in time ((Brav et al., 2008), (Klein and Zur, 2009)). In Germany, this was the case with The Children's Investment Fund's purchase of Deutsche Boerse shares in 2005 and M2 Capital's (now MarCap Investors) investment in

Cewe Color Holding in 2006. Cevian Capital's strategy was not to ask for changes to the dividend policy since there was apparently no pressure to create short-term liquidity for the investor.

IV.4.2. Supervisory Board Representation

The supervisory board structure before Cevian Capital's investment can be described as follows. According to Demag Cranes' articles of incorporation the number of supervisory board seats was twelve at the time of Cevian Capital's investment with one half of the seats being filled with shareholder representatives. The other half comprised labour representatives according to section 8 (1) of the articles of incorporation dated 20 March 2007. The term of office of the members of the supervisory board was five years, equalling the maximum allowed term of office. The last annual meeting to vote upon supervisory board appointments was held in 2008 as can be seen from the agenda of the respective annual meeting held on 06 March 2008. The next orderly supervisory board election would not take place until 2013.

The attendance rate at the annual meeting of Demag Cranes in March 2010, prior to the investment of Cevian Capital, was 45.21 percent according to a release by Demag Cranes AG dated 02 March 2010. The attendance rate at the 2011 annual meeting was slightly higher at 48.02 percent. This is below the average attendance rate of 52.1 percent calculated from DSW e.V. data between 1998 and 2008 for the companies listed in the DAX-30 index. The DAX-30 index is a German benchmark index comparable to the Dow Jones Industrial Average index. The lower attendance rate in this case is not necessarily surprising since Demag Cranes is a corporation with no major shareholders and a free float of close to 100 percent. Given an attendance rate of approximately 45 percent, a block of 22.5 percent of voting rights would be needed to win supervisory board seats in case of an election.

On 18 October 2010 London-based Centaurus Capital, a private investment company and event-driven equity investor, disclosed a 3.37 percent stake in Demag Cranes. In the meantime Allianz Global Investors, the asset management subsidiary of German insurance company Allianz Group, had become the only other investor to disclose a stake (4.33 percent) above regulatory thresholds.

On 22 October 2010, Demag Cranes released a press statement that supervisory board member Martin Posth had stepped down from office and the vacancy would be filled by Jens Tischendorf, a partner at Cevian Capital responsible for the fund's investments in Germany. This indicates that the supervisory board representation was not obtained through hostility but through communication with other shareholders and the supervisory board. Cevian Capital communicated its ambition to obtain interest in obtaining a board representation in its 27a

WpHG filing in May 2010. The investor then remained patient for six months before being assigned a board seat. Cevian Capital's approach was consensus-oriented and it accelerated the process of gaining supervisory board representation. The investor was not required to wait until the next orderly supervisory board election to be held in 2013.

Technically, if a board member steps down before the end of his term and the board has fewer than the required number of board members, a successor will be appointed by the local court of justice on an application by either the supervisory board itself, the management board or a shareholder (section 104 subsection 1 AktG). Article 5.4.3 DCGK provides that appointments by the local court should only be valid until the next annual meeting. Accordingly, there was a formal vote at the following annual meeting of Demag Cranes in March 2011 on the nomination of Jens Tischendorf to the supervisory board and it was passed with 99.37 percent of votes cast. This unanimous result suggests that consensus-oriented shareholder activism was supported by fellow shareholders.

Cevian Capital held more than 5 percent of voting rights. The investor could have forced a supervisory board election before the 2013 annual meeting (section 122 sub-section 2 AktG). However, as Hopt (1997) points out, a resolution to dismiss a member of the supervisory board would not require 50 but 75 percent of votes cast at the annual meeting (section 103 sub-section 1 AktG), provided there is no different threshold set out in the articles of incorporation. The articles of incorporation of Demag Cranes do not foresee a lower threshold.

Proceedings leading to a dismissal of a member of the supervisory board are a rare event in German corporate governance given this 75-percent requirement. Communication and consensus among shareholders in the Demag case was a viable path for obtaining board representation.

IV.4.3. Takeover by Terex Corporation

On 08 October 2010, the management board of Demag Cranes issued a press release commenting on alleged takeover plans by competitors. Demag confirmed the receipt of "highly preliminary and conditional indications of interest". The identity of potential acquirers was not disclosed. At the same time, newspapers reported that Konecranes, a competitor from Finland, and Terex Corporation had signalled interest in acquiring Demag Cranes. No formal takeover bid was launched. Neither before nor after Cevian Capital had obtained board representation an official statement with respect to the takeover rumours was issued from Cevian Capital – who was at that time the single largest shareholder of the firm. However, in February 2011, Centaurus Capital publicly urged Demag Cranes to start merger talks. A regulatory disclosure dated 03 January

2011 stated that Centaurus Capital had previously increased its stake to 5.06 percent.

Terex Corporation, which would later turn out to be the successful bidder, is a U.S. manufacturer of heavy equipment also active in the field of industrial cranes. In 2002, Terex had bought Demag Mobile Cranes from Siemens. This means the firm was acquainted with the Demag brand and with acquiring a German company. Terex announced its tender offer in May 2011. By 16 August 2011, 81.83 percent of the outstanding shares were tendered and ultimately transferred to Terex at an increased offer price.

When Terex Corporation announced the tender offer, Cevian Capital for the first time issued a public statement with respect to a potential takeover calling the offer at EUR 41.75 per share "inadequate". Cevian Capital eventually accepted the increased offer of EUR 45.50 per share. Demag Cranes chief executive officer (Vorstandsvorsitzender) Aloysius Rauen called this offer "fair and adequate". Rauen held approximately 0.02 percent of the shares of Demag Cranes. According to the annual report 2009 the shares were purchased in August 2009 and therefore not part of a compensation plan initiated by the supervisory board or Cevian Capital. Nevertheless, Cevian Capital is said to be an investor that actively ensures managers' interests are aligned with those of shareholders by promoting management share ownership (Sunesson, 2008). According to agency theory, the alignment of interests can reduce agency costs. Rauen's term of office as member of the management board was extended for another five years until 30 April 2017 by the supervisory board on 06 May 2011.

IV.4.4. Lessons Learned From the Demag Case

The Demag case evidence supports the strong shareholder rights perspective. Cevian Capital through its shareholder rights obtained supervisory board representation and actively participated in the firm's corporate governance. Furthermore, the transaction created substantial value for the investor. Assuming that the shares were purchased at an average market price of EUR 26 and sold at the increased offer price of EUR 45.50 per share this represents a gain of approximately 75 percent or EUR 41.5 million for the investor within one year. Demag's shares later peaked at EUR 60 per share and were trading above EUR 50 for most of the year 2012.

There is no information on whether or not Cevian Capital actively promoted a sale of the company before Terex Corporation announced its acquisition plans. There was generally a growing pressure for consolidation in the industrial cranes sector. It rather appears that the activist recognized that Demag Cranes was an undervalued company. This was stated in press statements accompanying the section 27a WpHG disclosure. Cevian Capital as well as Centaurus Capital took on the role of a catalyst in the takeover process. This shows how international investors participate in German corporate governance and how they and also other shareholders benefit from the actions taken.

Recent regulatory reforms of corporate law and securities law have created a more favourable environment for shareholder activism and shareholder monitoring. It appears that in Germany both the corporate governance framework of stock-market traded corporations and minority shareholder rights are strong. This is in line with the findings of Schaefer (2007) and Goergen, Manjon and Renneboog (2008).

Overall, the results do not create the view of a "strong" or "stronger" shareholder rights environment in the United States opposed to a "weak" or "weaker" environment in Germany. Rather it seems that some evidence points in the opposite direction. The comparison of shareholder rights with respect to the U.S. board of directors and the German supervisory board (Table IV.22.) support this finding

IV.5. Cevian Capital's Follow-up Investment in Bilfinger Berger SE

On 31 October 2011, a regulatory filing was issued by Bilfinger Berger SE disclosing a stake of 12.62 percent of voting rights held by Cevian Capital. The statement according to section 27a WpHG revealed intentions that were identical to the section-27a-statement issued in relation with the investment in Demag Cranes. Within four months of Cevian's investment the stock of Bilfinger Berger rose approximately 25 percent. This was in line, however, with the DAX-30 Index.

After Cevian Capital's investments in Daimler, Munich Re and Demag Cranes, its acquisition of a stake in Bilfinger Berger was the investor's fourth minority investment on the German market within the past five years. The following paragraph introduces Cevian Capital's new investment and draws conclusions with respect to the nature of the German corporate governance system.

Bilfinger Berger, originally one of the largest German construction companies, has changed its business model to become an international services group for industry, real estate and infrastructure. The firm was profitable before, during and after the financial crisis of 2008. According to the annual report 2010 the turnover for the full year was approximately EUR 8.0 billion.

In 2010, the firm decided to change its legal form from German Aktiengesellschaft (AG) to Societas Europaea (SE). The main purpose of the change of legal form to Societas Europaea is to simplify cross-border activities for large corporations within the European Union, reduce bureaucratic obstacles and enhance efficiency (Thoma and Leuering, 2002). There happens to be an increasing number of German corporations undertaking this step. From a shareholder rights perspective the change of legal form has in most cases a negligible effect. The German stock corporation act is mandatory law and it still applies (compare section 3 and section 20 of the Statute for a European Company). Bilfinger Berger did not change its two-tier corporate governance structure consisting of supervisory board and management board as a consequence of the change of legal form.

At the 2011 annual meeting a new supervisory board was elected for a full term of five years. The supervisory board has twelve members with half of the members being shareholder representatives. A change took place on the *management* board of Bilfinger Berger during the year 2011. Herbert Bodner stepped down and Roland Koch was appointed chief executive officer (Vorstandsvorsitzender).

Parallels exist between Cevian Capital's investment in Demag Cranes and its follow-up investment in Bilfinger Berger. Cevian Capital apparently remains an active minority shareholder in Germany. This further supports the strong shareholder rights perspective in Germany. If minority shareholder rights were weak or insufficient, Cevian Capital as a professional, international investor with the expertise to move to other markets would retreat from the German market. Cevian Capital would most likely *not* have made a follow-up investment within such a short period of time. The German corporate governance framework is apparently well-developed with respect to minority shareholder rights.

IV.6. Shareholder Activists on German Supervisory Boards

Is it at all possible for a minority shareholder of a German corporation to gain supervisory board representation even if he is not a majority owner of the corporation? According to the Demag case it is. Additionally, Table IV.24. provides an overview of more cases of supervisory board representation by activist minority shareholders. The table presents cases of *supervisory* board representation by activist shareholders. In the cases of Kuka, Deutsche Boerse, and Curanum there were also changes on the *management* board level attributable to activist minority shareholders.

Although in most cases the board seats were obtained at corporations without a large blockholder, the cases of Deutsche Telekom and Cewe Color Holding show that board seats can be obtained *despite the existence of a large shareholder.*

Schaefer (2007) describes supervisory board representation as the most active form of shareholder activism. He remarks that recent reforms of the regulatory framework have created a favourable environment for shareholder activism in Germany. He further points out that the rules and regulations providing the legal environment for active minority shareholders appear to be well-balanced also from the management's point of view. Schaefer's study does not support the weak shareholder rights view of La Porta et al. ((1998), (2000a)) and Kim et al. (2007) either.

Little evidence exists for shareholder activist events before the year 2000. Only two studies by Jenkinson and Ljungqvist (2001) and Croci (2007) present empirical evidence for the period prior to the year 2000. Consistent with the view of La Porta et al. shareholder activism did not have a meaningful impact on German corporations before that time (Steiger, 2000).

Since the influence of banks has decreased in the past two decades ((Vitols, 2005), (Dittmann et al., 2010)) and regulatory initiatives have increased accountability and transparency in corporate Germany ((Nowak, 2004), (Goergen et al., 2008)), the environment for small shareholders has changed. More recent research on minority shareholder activism in Germany documents substantial activity by international, professional minority investors ((Bessler et al., 2010), (Mietzner et al., 2011), (Drees et al., 2011)). Assuming these investors are well-educated capital market participants they would not invest in a country whose governance system is lacking small shareholder protection.

Table IV.24. *Shareholder Activists on German Supervisory Boards:* The table shows cases where one or more supervisory board seats have been obtained through shareholder activism. Investor / Fund Manager is the perceived leading activist among all activists. Max. %-stake is the highest percentage of voting rights on record held by the activist. Year stake(s) is the calendar year when the activist disclosed its stake for the first time. Year seat(s) is the calendar year when the activist obtained board representation. Board representation can be obtained either by the activist directly or by a delegate. EURm target market cap. is the target firm's market capitalisation at the end of the quarter preceding the activist's investment. Major corporate events lists important events that occurred if the activist was still invested at the respective event date. Cases where activist minority shareholders have not been successful in obtaining board seats include Babcock Borsig AG (2002), Volkswagen AG (2006), Cewe Color Holding AG (2007) and Ehlebracht AG (2010). Source: Lexis Nexis database, target annual reports, regulatory WpHG filings, target websites. *In 2012, more than half of the shares of Augusta Technologie AG were acquired by a third party

Target firm	Investor / Fund Manager	Max. %-stake	Year stake / Year seat(s)	Way of obtaining board seat(s)	EURm target market cap.	Major corporate events
Kuka AG (Iwka AG)	Guy Wyser Pratte (Wyser Pratte Mgmt Co.)	9.7	2003 / 2009	Appointment by local court after member steps down	426	Series of divestitures
Comtrade AG	Florian Homm (FM Fund Management Ltd.)	29.6	2004 / 2004	Extraordinary general meeting	24	Bankruptcy
Deutsche Boerse AG	Christopher Hohn (TCI Fund Management LLP)	10.1	2005 / 2006	Chairman of the supervisory board steps down from office	4,947	Takeover of LSE blocked
Cewe Color Holding AG	David Marcus (M2 Capital Mgmt L.P.)	10.3	2005 / 2006	Appointment according to section 101 sub-section 2 AktG	232	-
Euromicron AG	Longview S.A. (Sapinda Intl. Ltd.)	11.3	2005 / 2006	Appointment by local court after member steps down	84	-

Deutsche Telekom AG	The Blackstone Group	4.5	2006 / 2006	Appointment by local court after member steps down	58,410	-
Techem AG	Elliott Capital Advisors LLP	15.2	2006 / 2007	Elected at annual meeting	1,360	Target firm sold
Augusta Technologie AG	Lincoln Vale LLP	17.1	2007 / 2010	Dismissal of incumbent board at annual meeting and new appointment	101	-*
Curanum AG	Audley Capital Management Ltd.	12.8	2007 / 2009	Member steps down after proposal for dismissal followed by election of new member at annual meet.	297	-
Sky Deutschland AG	Centaurus Capital Ltd.	3.1	2008 / 2009	Elected at annual meeting	1,573	-
Pulsion Medical Systems AG	Shareholder Value Beteiligungen AG	5.0	2009 / 2009	Nominated and elected at annual meeting	19	-
Demag Cranes AG	Cevian Capital LLP	10.1	2010 / 2010	Appointment by local court after member steps down	500	Target firm sold
Infineon Technologies AG	Hermes Focus Asset Management	0.5	2009 / 2011	Chairman of the supervisory board steps down from office	2,935	-

IV.7. Conclusion

The United States and Germany have diverging legal systems and rules governing stock corporations. This analysis is a discussion of these differences in the corporate governance framework with a focus on the German supervisory board. Although there is some convergence between the two systems of corporate law as Hellgardt and Hoger (2011) point out, in practice, differences such as the one-tier corporate governance system in the U.S. and the two-tier system in Germany do remain. The influence of banks is declining and regulatory initiatives in Germany have created transparency for investors. The weak shareholder rights view –with small investors not participating in corporate governance– is outdated. The findings by La Porta, Lopez-de-Silanes, Shleifer, and Vishny with respect to the German market need to be reviewed. Small- and medium-sized firms should be included in any sample to avoid a bias towards large firms.

The case of activist shareholder Cevian Capital acquiring a 10.07 percent stake in German stock corporation Demag Cranes in May 2010 followed by the acquisition of Demag Cranes by U.S. competitor Terex Corporation presented some evidence for the strong shareholder rights perspective. There are three key takeaways from the Demag Cranes case: (i) minority shareholder rights in Germany appear to be strong, (ii) minority shareholders do have the possibility to obtain board representation, and (iii) minority shareholder activism can have an impact on a firm's corporate governance.

The evidence presented in this study further includes an overview of cases of supervisory board representation by small investors. Overall, the empirical evidence supports the strong shareholder rights perspective in Germany, with small investors actively participating in corporate governance.

V. Summary of Key Research Findings

1. Activity by shareholder activists has increased during the past decade and minority shareholders actively participate in corporate governance. These results are in line with the most recent research on shareholder activism in Germany but conflict with the weak shareholder rights perspective and the stereotypical view of German corporate governance of the past century.
2. Shareholder activism leads to a significant, positive increase in shareholder value as measured by a short-term event study. This confirms the results of earlier studies.
3. The magnitude of abnormal returns apparently depends on the credibility of the activist effort. A *New Supervisory Board Election Timing (NewBET)* variable was introduced for the first time and applied in a multivariate regression model. An activist approach appears to be less credible if it is not well-timed and if there is a large cash position on the target's balance sheet.
4. Following the investment of potential shareholder activists, the attendance rate at the annual meeting increases significantly. This new finding suggests that the activists actively participate in corporate governance through monitoring.
5. The theoretical potential for influencing German corporations is substantial, especially given the low annual meeting attendance rates observed in the sample. However, the empirical evidence based on publicly available information suggests that the overall level of actual influence is rather moderate. This partly confirms prior research findings.
6. The weak shareholder rights perspective needs to be further reviewed given regulatory and factual changes to the German corporate governance system.

References

Achleitner, A.-K., Andres, C., Betzer, A., and Weir, C. (2011). Wealth effects of private equity investments on the German stock market. The European Journal of Finance, 17, 217-239.

Achleitner, A.-K., Betzer, A., and Gider, J. (2010). Do Corporate Governance Motives Drive Hedge Fund and Private Equity Fund Activities? European Financial Management, 16, 805-828.

Andres, C. (2008). Large shareholders and firm performance - An empirical examination of founding-family ownership. Journal of Corporate Finance, 14, 431-445.

Auer, L. v. (2007). Ökonometrie. Berlin, Heidelberg: Springer.

Bassen, A. (2002). Institutionelle Investoren und Corporate Governance. Wiesbaden: Deutscher Universitäts-Verlag, Gabler.

Bassen, A., Königs, A., and Schiereck, D. (2008). Aktionärsstrukturabhängige Investor Relations bei deutschen Großunternehmen. Zeitschrift für Corporate Governance, 3, 101-105.

Baums, T., Drinhausen, F., and Keinath, A. (2011). Anfechtungsklagen und Freigabeverfahren. Eine empirische Studie. ZIP Zeitschrift für Wirtschaftsrecht, 49, 2329-2352.

Baums, T., Keinath, A., and Gajek, D. (2007). Fortschritte bei Klagen gegen Hauptversammlungsbeschlüsse? Eine empirische Studie. ZIP Zeitschrift für Wirtschaftsrecht, 35, 1629-1650.

Bebchuk, L. A. (2005). The case for increasing shareholder power. Harvard Law Review, 118, 833-914.

Bebchuk, L. A. (2007). The Myth of the Shareholder Franchise. Virginia Law Review, 93, 675-732.

Bebchuk, L. A., Coates IV, J. C., and Subramanian, G. (2002a). The Powerful Antitakeover Force of Staggered Boards: Further Findings and a Reply to Symposium Participants. *Stanford Law Review, 55*, 885-917.

Bebchuk, L. A., Coates IV, J. C., and Subramanian, G. (2002b). The Powerful Antitakeover Force of Staggered Boards: Theory, Evidence, and Policy. *Stanford Law Review, 54*, 887-951.

Bebchuk, L. A., and Cohen, A. (2005). The cost of entrenched boards. *Journal of Financial Economics, 78*, 409-433.

Becht, M. (1997). Strong Blockholders, Weak Owners and the Need for European Mandatory Disclosure. *European Corporate Governance Network Executive Report*, 1-118.

Becht, M., and Boehmer, E. (2003). Voting control in German corporations. *International Review of Law and Economics, 23*, 1-29.

Becht, M., Franks, J., and Grant, J. (2010a). Hedge Fund Activism in Europe. *Finance Working Paper No. 283/2010, European Corporate Governance Institute*.

Becht, M., Franks, J., Mayer, C., and Rossi, S. (2010b). Returns to Shareholder Activism: Evidence from a Clinical Study of the Hermes UK Focus Fund. *Review of Financial Studies, 23*, 3093-3129.

Berle, A. A., and Means, G. C. (1932). The Modern Corporation and Private Property. New York: Macmillan.

Bessler, W., Drobetz, W., and Holler, J. (2008). Capital Markets and Corporate Control: Empirical Evidence from Hedge Fund Activism in Germany. *Working Paper, Justus-Liebig-Universität, Gießen*.

Bessler, W., Drobetz, W., and Holler, J. (2010). The Returns to Hedge Fund Activism in Germany. *Working Paper presented at the 2011 Annual Meeting of the Midwest Finance Association, Chicago, Illinois*.

Bethel, J. E., Liebeskind, J. P., and Opler, T. (1998). Block Share Purchases and Corporate Performance. *Journal of Finance, 53*, 605-634.

Bizjak, J. M., and Marquette, C. J. (1998). Are Shareholder Proposals All Bark and No Bite? Evidence from Shareholder Resolutions to Rescind Poison Pills. *Journal of Financial and Quantitative Analysis, 33*, 499-521.

Boehmer, E., Musumeci, J., and Poulsen, A. B. (1991). Event-study methodology under conditions of event-induced variance. *Journal of Financial Economics, 30*, 253-272.

Brass, S. (2010). Hedgefonds als aktive Investoren. Frankfurt/ Main: Peter Lang.

Brav, A., Jiang, W., Partnoy, F., and Thomas, R. (2008). Hedge Fund Activism, Corporate Governance, and Firm Performance. *Journal of Finance, 63*, 1729-1775.

Brown, S. J., and Warner, J. B. (1980). Measuring Security Price Performance. *Journal of Financial Economics, 8*, 205-258.

Brown, S. J., and Warner, J. B. (1985). Using Daily Stock Returns: The Case of Event Studies. *Journal of Financial Economics, 14*, 3-31.

Carleton, W. T., Nelson, J. M., and Weisbach, M. S. (1998). The Influence of Institutions on Corporate Governance through Private Negotiations: Evidence from TIAA-CREF. *Journal of Finance, 53*, 1335-1362.

Cary, W. L. (1974). Federalism and Corporate Law: Reflections Upon Delaware. *The Yale Law Journal, 83*, 663.

Corrado, C. J. (2011). Event studies: A methodology review. *Accounting and Finance, 51*, 207-234.

Croci, E. (2007). Corporate Raiders, Performance and Governance in Europe. *European Financial Management, 13*, 949-978.

Cziraki, P., Renneboog, L., and Szilagyi, P. G. (2010). Shareholder Activism through Proxy Proposals: The European Perspective. *European Financial Management, 16*, 738-777.

Dauner-Lieb, B. (2007). Aktuelle Vorschläge zur Präsenzsteigerung in der Hauptversammlung. *Zeitschrift für Wirtschafts- und Bankrecht (Wertpapier-Mitteilungen), 61*, 9-17.

Dittmann, I., Maug, E., and Schneider, C. (2010). Bankers on the Boards of German Firms: What They Do, What They Are Worth, and Why They Are (Still) There. *Review of Finance, 14*, 35-71.

Dixon, W. J. (1960). Simplified Estimation from Censored Normal Samples. *The Annals of Mathematical Statistics, 31*, 385-391.

Doh, J. P., and Guay, T. R. (2006). Corporate Social Responsibility, Public Policy, and NGO Activism in Europe and the United States: An Institutional-Stakeholder Perspective. *Journal of Management Studies, 43*, 47-73.

Donaldson, L., and Davis, J. H. (1991). Stewardship Theory or Agency Theory: CEO Governance and Shareholder Returns. *Australian Journal of Management, 16*, 49-64.

Dooley, M. P., and Goldman, M. D. (2001). Some Comparisons Between the Model Business Corporation Act and the Delaware General Corporation Law. *The Business Lawyer, 56*, 737-766.

Drees, F., Mietzner, M., and Schiereck, D. (2011). New minority blockholders, performance and governance in Germany. *Working Paper, European Business School, Oestrich-Winkel.*

Drerup, T. (2010). Much Ado about Nothing. The Effects of Hedge Fund Activism in Germany. *Working Paper, University of Bonn.*

Drobetz, W., Schillhofer, A., and Zimmermann, H. (2004). Corporate Governance and Expected Stock Returns: Evidence from Germany. *European Financial Management, 10*, 267-293.

DSW e.V. (2008). Attendance rates at the annual meetings of the DAX-30 companies (1998-2008) in percent of voting rights. Düsseldorf: DSW e.V.

Ernst, E., Gassens, J., and Pellens, B. (2009). Verhalten und Präferenzen deutscher Aktionäre. Frankfurt am Main: Deutsches Aktieninstitut e.V.

Fama, E. F. (1970). Efficient Capital Markets: A Review of Theory and Empirical Work. *Journal of Finance, 25*, 383-417.

Fama, E. F. (1998). Market efficiency, long-term returns, and behavioral finance. *Journal of Financial Economics, 49*, 283-306.

Fama, E. F., and Jensen, M. C. (1983). Separation of Ownership and Control. *Journal of Law and Economics, 26*, 301-325.

Fauver, L., and Fuerst, M. E. (2006). Does good corporate governance include employee representation? Evidence from German corporate boards. *Journal of Financial Economics, 82*, 673-710.

Fleischer, H. (2008). Finanzinvestoren im ordnungspolitischen Gesamtgefüge von Aktien-, Bankaufsichts- und Kapitalmarktrecht. *ZGR Zeitschrift für Unternehmens- und Gesellschaftsrecht, 37*, 185-224.

Franks, J., and Mayer, C. (1998). Bank control, takeovers and corporate governance in Germany. *Journal of Banking & Finance, 22*, 1385-1403.

Franks, J., and Mayer, C. (2001). Ownership and Control of German Corporations. *Review of Financial Studies, 14*, 943-977.

Gillan, S. L., and Starks, L. T. (2000). Corporate governance proposals and shareholder activism: the role of institutional investors. *Journal of Financial Economics, 57*, 275-305.

Gillan, S. L., and Starks, L. T. (2007). The Evolution of Shareholder Activism in the United States. *Journal of Applied Corporate Finance, 19*, 55-73.

Goergen, M., Manjon, M. C., and Renneboog, L. (2008). Recent developments in German corporate governance. *International Review of Law and Economics, 28*, 175-193.

Gompers, P., Ishii, J., and Metrick, A. (2003). Corporate Governance and Equity Prices. *Quarterly Journal of Economics, 118*, 107-188.

Gorton, G., and Schmid, F. A. (2004). Capital, labor, and the firm: a study of German codetermination. *Journal of the European Economic Association, 2*, 863-905.

Gospel, H., Haves, J., Vitols, S., Voss, E., and Wilke, P. (2009). The impacts of private equity investors, hedge funds and sovereign wealth funds on industrial restructuring in Europe as illustrated by case studies (report on behalf of the European Commission). Hamburg.

Greenwood, R., and Schor, M. (2009). Investor activism and takeovers. *Journal of Financial Economics, 92*, 362-375.

Hackethal, A., Schmidt, R. H., and Tyrell, M. (2005). Banks and German Corporate Governance: on the way to a capital market-based system? *Corporate Governance: An International Review, 13*, 397-407.

Heiss, F., and Köke, J. (2004). Dynamics in Ownership and Firm Survival: Evidence from Corporate Germany. *European Financial Management, 10*, 167-195.

Hellgardt, A., and Hoger, A. (2011). Transatlantische Konvergenz der Aktionärsrechte - Systemvergleich und neuere Entwicklungen. *ZGR Zeitschrift für Unternehmens- und Gesellschaftsrecht, 40*, 38-82.

Helwege, J., Intintoli, V. J., and Zhang, A. (2012). Voting with their feet or activism? Institutional investors' impact on CEO turnover. *Journal of Corporate Finance, 18*, 22-37.

Hofstetter, K. (2008). Von der "Landsgemeinde"- zur "proxy"-Generalversammlung: Vorschläge für einen Paradigmenwechsel in der Schweiz. *ZGR Zeitschrift für Unternehmens- und Gesellschaftsrecht, 37*, 560-592.

Homberg, F., and Osterloh, M. (2010). Fusionen und Übernahmen im Licht der Hybris - Überblick über den Forschungsstand. *Journal für Betriebswirtschaft, 60*, 269-294.

Hopt, K. J. (1997). The German Two-Tier Board (Aufsichtsrat) - A German View on Corporate Governance. In Hopt & Wymeersch (Eds.), *Comparative Corporate Governance: Essays and Materials*. Berlin, New York: de Gruyter.

Hopt, K. J., and Leyens, P. C. (2004). Board Models in Europe. Recent Developments of Internal Corporate Governance Structures in Germany, the United Kingdom, France, and Italy. *Law Working Paper No. 18/2004, European Corporate Governance Institute*.

Iber, B. (1985). Zur Entwicklung der Aktionärsstruktur in der Bundesrepublik Deutschland (1963-1983). *Zeitschrift für Betriebswirtschaft, 55*, 1101-1119.

Jenkinson, T., and Ljungqvist, A. (2001). The role of hostile stakes in German corporate governance. *Journal of Corporate Finance, 7*, 397-446.

Jensen, M. C., and Meckling, W. H. (1976). Theory of the Firm: Managerial Behavior, Agency Costs and Ownership Structure. *Journal of Financial Economics, 3*, 305-360.

Kahan, M., and Rock, E. B. (2007). Hedge Funds in Corporate Governance and Corporate Control. *University of Pennsylvania Law Review, 155*, 1021-1093.

Karpoff, J. M. (2001). The Impact of Shareholder Activism on Target Companies: A Survey of Empirical Findings. *Working Paper, University of Washington*.

Kaserer, C., Achleitner, A.-K., Einem, C. v., and Schiereck, D. (2007). Private Equity in Deutschland: Rahmenbedingungen, ökonomische Bedeutung und Handlungsempfehlung. Norderstedt: Books on Demand.

Kim, K. A., Kitsabunnarat-Chatjuthamard, P., and Nofsinger, J. R. (2007). Large shareholders, board independence, and minority shareholder rights: Evidence from Europe. *Journal of Corporate Finance, 13*, 859-880.

Kirk III, W. E. (1984). A Case Study in Legislative Opportunism: How Delaware Used the Federal-State System to Attain Corporate Pre-Eminence. *Journal of Corporation Law, 10*, 233-260.

Klein, A., and Zur, E. (2009). Entrepreneurial Shareholder Activists: Hedge Funds and Other Private Investors. *Journal of Finance, 64*, 187-229.

Köke, J. (2004). The market for corporate control in a bank-based economy: a governance device? *Journal of Corporate Finance, 10*, 53-80.

La Porta, R., Lopez-de-Silanes, F., and Shleifer, A. (1999). Corporate Ownership Around the World. *Journal of Finance, 54*, 471-517.

La Porta, R., Lopez-de-Silanes, F., and Shleifer, A. (2006). What Works in Securities Laws? *Journal of Finance, 61*, 1-32.

La Porta, R., Lopez-de-Silanes, F., Shleifer, A., and Vishny, R. W. (1997). Legal Determinants of External Finance. *Journal of Finance, 52*, 1131-1150.

La Porta, R., Lopez-de-Silanes, F., Shleifer, A., and Vishny, R. W. (1998). Law and Finance. *Journal of Political Economy, 106*, 1113-1155.

La Porta, R., Lopez-de-Silanes, F., Shleifer, A., and Vishny, R. W. (2000a). Agency Problems and Dividend Policies around the World. *Journal of Finance, 55*, 1-33.

La Porta, R., Lopez-de-Silanes, F., Shleifer, A., and Vishny, R. W. (2000b). Investor protection and corporate governance. *Journal of Financial Economics, 58*, 3-27.

La Porta, R., Lopez-de-Silanes, F., Shleifer, A., and Vishny, R. W. (2002). Investor Protection and Corporate Valuation. *Journal of Finance, 57*, 1147-1170.

Levit, D., and Malenko, N. (2011). Nonbinding Voting for Shareholder Proposals. *Journal of Finance, 66*, 1579-1614.

MacKinlay, A. C. (1997). Event Studies in Economics and Finance. *Journal of Economic Literature, 35*, 13-39.

Mietzner, M., and Schweizer, D. (2007). Hedge Funds vs. Private Equity Funds as Shareholder Activists - Differences in Value Creation. *Working Paper, European Business School, Oestrich-Winkel.*

Mietzner, M., Schweizer, D., and Tyrell, M. (2011). Intra-Industry Effects of Shareholder Activism in Germany - Is There a Difference between Hedge Fund and Private Equity Investments? *Schmalenbach Business Review, 63*, 151-185.

Moeller, S. B., Schlingemann, F. P., and Stulz, R. M. (2004). Firm size and the gains from acqusitions. *Journal of Financial Economics, 73*, 201-228.

Moeller, S. B., Schlingemann, F. P., and Stulz, R. M. (2005). Wealth Destruction on a Massive Scale? A Study of Acquiring-Firm Returns in the Recent Merger Wave. *Journal of Finance, 60*, 757-782.

Nowak, E. (2004). Investor Protection and Capital Market Regulation in Germany. In J. P. Krahnen & R. H. Schmidt (Eds.), *The German Financial System*. New York: Oxford University Press.

Parrino, R., Sias, R. W., and Starks, L. T. (2003). Voting with their feet: institutional ownership changes around forced CEO turnover. *Journal of Financial Economics, 68*, 3-46.

Prevost, A. K., Rao, R. P., and Williams, M. A. (2012). Labor Unions as Shareholder Activists: Champions or Detractors? *The Financial Review, 47*, 327-349.

Prigge, S., and Steenbock, A. (2002). Pensionsfondsaktivismus - Chancen und Risiken einer Reform der Alterssicherung für die Corporate Governance in Deutschland. *Zeitschrift für Betriebswirtschaft, 72*, 777-795.

Rauch, C., and Umber, M. (2012). Private Equity Shareholder Activism. *Working Paper, Goethe University, Frankfurt am Main.*

Schaefer, H. (2007). Shareholder Activism und Corporate Governance. *Neue Zeitschrift für Gesellschaftsrecht, 10*, 900-903.

Schäfer, H., and Hertrich, C. (2011). Shareholder Activism in Germany - Theoretical Considerations and Empirical Evidence. *Working paper, University of Stuttgart.*

Schiereck, D., and Steiger, M. (2001). Internationale Corporate Governance durch institutionelle Anleger. In S. A. Jansen, G. Picot & D. Schiereck (Eds.), *Internationales Fusionsmanagement.* Stuttgart: Schaeffer-Poeschel.

Schiereck, D., and Thamm, C. (2012). Der Squeeze-out bei der Teutonia Zementwerk AG - eine Fallstudie zur Aktivität der Berufskläger. *Zeitschrift für Corporate Governance, 9,* 211-216.

Schmolke, K. U. (2007). Institutionelle Anleger und Corporate Governance - Traditionelle institutionelle Investoren vs. Hedgefonds. *ZGR Zeitschrift für Unternehmens- und Gesellschaftsrecht, 5,* 701-744.

Shleifer, A., Morck, R., and Vishny, R. W. (1990). Do Managerial Objectives Drive Bad Acquisitions? *Journal of Finance, 45,* 31-48.

Shleifer, A., and Vishny, R. W. (1989). Management Entrenchment: The Case of Manager-Specific Investments. *Journal of Financial Economics, 25,* 123-140.

Shleifer, A., and Vishny, R. W. (1997). A Survey of Corporate Governance. *The Journal of Finance, 52,* 737-783.

Smith, A. (1776). An Inquiry into the Nature and Causes of the Wealth of Nations. Oxford: Clarendon Press.

Stadler, M. (2010). Shareholder-Aktivismus durch Hedge Fonds. Dissertation: Technische Universität Berlin.

Steiger, M. (2000). Institutionelle Investoren im Spannungsfeld zwischen Aktienmarktliquidität und Corporate Governance. Baden-Baden: Nomos.

Sunesson, D. (2008). Ownership Matters: A Clinical Study of Cevian Capital. *Working Paper, Stockholm School of Economics, Stockholm.*

Thoma, G. F., and Leuering, D. (2002). Die Europäische Aktiengesellschaft - Societas Europaea. *NJW Neue Juristische Wochenschrift, 55,* 1449-1458.

Tirole, J. (2006). The Theory of Corporate Finance. Princeton and Oxford: Princeton University Press.

Unknown Author. (1961). Presentation of court decision by the German national supreme court Bundesgerichtshof dated 23 November 1961 (II ZR 4/60). *Zeitschrift für Wirtschafts- und Bankrecht (Wertpapier-Mitteilungen), 15,* 1324.

Unknown Author. (1962). "Pression" eines Minderheitsaktionärs durch Anfechtungsklage? *Betriebs-Berater, 17,* 426.

Vitols, S. (2005). Changes in Germany's Bank-Based Financial System: implications for corporate governance. *Corporate Governance: An International Review, 13,* 386-396.

Weber, P., and Zimmermann, H. (2011). Hedge Fund Activism and Information Disclosure: the Case of Germany. *European Financial Management, Online version of record published before inclusion in an issue.*

Wenger, E. (2008). Die Geisterbahnfahrer von Bullshit Castle. In F. W. Wagner, T. Schildbach & D. Schneider (Eds.), *Private und öffentliche Rechnungslegung.* Wiesbaden: Gabler.

White, H. (1980). A Heteroskedasticity-Consistent Covariance Matrix Estimator and a Direct Test for Heteroskedasticity. *Econometrica, 48,* 817-838.

Wilcoxon, F. (1945). Individual Comparisons by Ranking Methods. *Biometrics Bulletin, 1,* 80-93.

Yermack, D. (2006). Flights of fancy: Corporate jets, CEO perquisites and inferior shareholder returns. *Journal of Financial Economics, 80,* 211-242.

Corporate Finance and Governance

Herausgegeben von Dirk Schiereck

Band 1   Sebastian Michael Gläsner: Return Patterns of German Open-End Real Estate Funds. An Empirical Explanation of Smooth Fund Returns. 2010.

Band 2   Patrick Ams: Directors' Dealings and Insider Trading in Germany. An Empirical Analysis. 2010.

Band 3   Joachim Vogt: Value Creation within the Construction Industry. A Study of Strategic Takeovers. 2011.

Band 4   Fabian Braemisch: Underpricing, Long-Run Performance, and Valuation of Initial Public Offerings. 2011.

Band 5   Matthäus Markus Sielecki: Creating and Governing an Integrated Market for Retail Banking Services in Europe. A Conceptual-Empirical Study of the Role of Regulation in Promoting a Single Euro Payments Area. 2011.

Band 6   Arne Wilkes: Determinants of Credit Spreads. An Empirical Analysis for the European Corporate Bond Market. 2011.

Band 7   Dirk Schiereck / Martin Setzer (Hrsg.): Bankerfolg und Akquisitionen. Aktuelle Erkenntnisse zur Konsolidierung und Restrukturierung der Finanzindustrie. 2011.

Band 8   Steffen Meinshausen: M&A Activity, Divestitures and Initial Public Offerings in the Fashion Industry. 2012.

Band 9   Malte Helmut Raudszus: Financial Return Risk and the Effect on Shareholder Wealth. 2012.

Band 10   Ramit Mehta: Mergers and Acquisitions in the Global Brewing Industry. A Capital Market Perspective. 2012.

Band 11   Marcel Normann: The Influence of German Top Executives on Corporate Policy and Firm Performance. 2012.

Band 12   Christoph Böhm: Risk-Adjusted Performance and Bank Governance Structures. 2013.

Band 13   Christian Thamm: Minority Shareholder Monitoring and German Corporate Governance. Empirical Evidence and Value Effects. 2013.

www.peterlang.de